The Palgrave Lacan Series

Series Editors
Calum Neill
Edinburgh Napier University
Edinburgh, UK

Derek Hook
Duquesne University
Pittsburgh, USA

Jacques Lacan is one of the most important and influential thinkers of the 20th century. The reach of this influence continues to grow as we settle into the 21st century, the resonance of Lacan's thought arguably only beginning now to be properly felt, both in terms of its application to clinical matters and in its application to a range of human activities and interests. The Palgrave Lacan Series is a book series for the best new writing in the Lacanian field, giving voice to the leading writers of a new generation of Lacanian thought. The series will comprise original monographs and thematic, multi-authored collections. The books in the series will explore aspects of Lacan's theory from new perspectives and with original insights. There will be books focused on particular areas of or issues in clinical work. There will be books focused on applying Lacanian theory to areas and issues beyond the clinic, to matters of society, politics, the arts and culture. Each book, whatever its particular concern, will work to expand our understanding of Lacan's theory and its value in the 21st century.

John Dall'Aglio
A Lacanian Neuropsychoanalysis
Consciousness Enjoying Uncertainty

John Dall'Aglio
Psychology
Duquesne University
Pittsburgh, PA, USA

ISSN 2946-4196 ISSN 2946-420X (electronic)
The Palgrave Lacan Series
ISBN 978-3-031-68830-0 ISBN 978-3-031-68831-7 (eBook)
https://doi.org/10.1007/978-3-031-68831-7

© The Author(s) 2024

This work is subject to copyright. All rights are solely and exclusively licensed by the Publisher, whether the whole or part of the material is concerned, specifically the rights of translation, reprinting, reuse of illustrations, recitation, broadcasting, reproduction on microfilms or in any other physical way, and transmission or information storage and retrieval, electronic adaptation, computer software, or by similar or dissimilar methodology now known or hereafter developed.
The use of general descriptive names, registered names, trademarks, service marks, etc. in this publication does not imply, even in the absence of a specific statement, that such names are exempt from the relevant protective laws and regulations and therefore free for general use.
The publisher, the authors and the editors are safe to assume that the advice and information in this book are believed to be true and accurate at the date of publication. Neither the publisher nor the authors or the editors give a warranty, expressed or implied, with respect to the material contained herein or for any errors or omissions that may have been made. The publisher remains neutral with regard to jurisdictional claims in published maps and institutional affiliations.

This Palgrave Macmillan imprint is published by the registered company Springer Nature Switzerland AG.
The registered company address is: Gewerbestrasse 11, 6330 Cham, Switzerland

If disposing of this product, please recycle the paper.

Foreword: The New Lacanian Brain

Jacques Lacan, at one point in his deservedly celebrated twentieth seminar (*Encore* [1972–1973]), offers a provocative aside about matters biological. Specifically, in the December 19, 1972, session of *Seminar XX*, he makes reference to deoxyribonucleic acid (DNA). This reference is in response to "a question... of the status of the notion of information whose success has been so lightening fast that one can say that the whole of science manages to get infiltrated by it."[1] After raising this issue thusly, Lacan states:

> We're at the level of the gene's molecular information and of the winding (*enroulements*) of nucleoproteins around strands of DNA, that are themselves wrapped (*enroulées*) around each other, all of that being tied together by hormonal links (*liens hormonaux*)—that is, messages that are sent, recorded (*s'enregistrent*), etc. Let us note that the success of this formula finds its indisputable source in a linguistics that is not only immanent but explicitly formulated (*bel et bien formulée*). In any case, this action extends right to the very foundations of scientific thought, being articulated as negative entropy (*néguentropie*).[2]

---

[1] Jacques Lacan, *The Seminar of Jacques Lacan, Book XX: Encore, 1972–1973* [ed. Jacques-Alain Miller; trans. Bruce Fink], New York: W.W. Norton and Company, 1998, pg. 17.

[2] Jacques Lacan, *Le Séminaire de Jacques Lacan, Livre XX: Encore, 1972–1973* [ed. Jacques-Alain Miller], Paris: Éditions du Seuil, 1975, pg. 22). (Lacan, *The Seminar of Jacques Lacan, Book XX*, pg. 17.

Making reference to a neologism, "linguistricks" (*linguisterie*),[3] he introduces at the start of this same seminar session, he then asks, "Is that what I, from another locus, that of my linguistricks, gather (*recueille*) when I make use of the function of the signifier?"[4] What is one to make of these reflections by Lacan concerning DNA, information, and language (whether as per linguistics or *à la* Lacanian *linguisterie*)? What is their significance?

Lacan is here registering how the natural sciences in general (i.e., "the whole of science") come to point in the direction of an information-theoretic metaphysics—with DNA serving him as an example of this especially fitting in the context of metapsychological discussions of (sexed/sexuated) bodies. Lacan's artful manipulation of Saussurean structural linguistics (i.e., his linguistricks) involves him broadening its theory of signifying systems and its models of the structural dynamics of signifiers to cover things well beyond human natural languages. This inflation of Saussurean linguistics into Lacanian *linguisterie* enables the latter to "gather" (*recueille* as collect or take in) under its heading not only the semantics and syntaxes of natural languages, but also units and relations within asubjective and non-human ontological regions too (including those regions forming the explanatory jurisdictions of the sciences of nature).

In fact, Lacan boldly is proposing on this 1972 occasion that the natural-scientific notion of information be more precisely conceptualized along the lines of his logic of the signifier. That is to say, Lacan's contention in this context is that even pre-human nature already contains within itself and is configured by its own symbolic orders, its self-organizing (as per negative entropy [*néguentropie*]) networks of differentiated elements and patterned connections constitutive of various sorts of information. A "linguistics" in Lacan's expanded *linguisterie*-sense is "immanent," as he claims, to nature writ large. Long pre-dating the appearance of the species *Homo sapiens* and the many languages developed by this species, there already are signifiers in the material Real. Or, put differently, the physical

---

[3] Lacan, *The Seminar of Jacques Lacan, Book XX*, pg. 15.
[4] Lacan, *The Seminar of Jacques Lacan, Book XX*, pg. 17.

universe involves not only a material Real, but also a material Symbolic (in the forms of various sorts of information). As DNA nicely illustrates for this Lacan, "messages… are sent, recorded, etc." within non-human nature—with this being articulated (i.e., "explicitly formulated") by natural scientists themselves (such as James Watson and Francis Crick).

This moment in the twentieth seminar during which Lacan invokes DNA is far from the only time when he dips into the domains of the life sciences specifically. However, Lacan's own various and often sophisticated considerations of biological topics have tended to be neglected by the majority of Lacanian thinkers and scholars. Many Lacanians have mistaken Lacan's justifiably harsh criticisms of the false naturalizations of Freudian metapsychology by non-Lacanian analysts engaging in scientistic misappropriations of bits and pieces of biology for a sweeping, unqualified dismissal of the life sciences in their entirety as utterly irrelevant to psychoanalysis. This mistake usually is an expression of an equally erroneous bigger picture of Lacan as an uncompromising anti-naturalist averse to any and every engagement with the empirical natural sciences.

With the neurobiology of the human organism as the natural-scientific sub-field most proximate (and, arguably, relevant) to the structures and phenomena characteristic of the psychical subjectivities at stake in psychoanalysis, a few brave souls have ventured, in recent years, to attempt certain kinds of Lacan-informed reckonings with the neurosciences. Such figures as François Ansermet, Ariane Bazan, Pierre Magistretti, Catherine Malabou, Catherine Morin, Arlette Pellé, Gérard Pommier, Edith Seifert, Mai Wegner, and Slavoj Žižek—this list is not exhaustive—each have addressed points of convergence and divergence between Lacanian analysis and biological investigations into the human central nervous system. Building on the labors of these pathbreaking pioneers, John Dall'Aglio has put together a formidably rigorous, detailed, and innovative conceptual vision of what would amount to a systematic Lacanian neuro-psychoanalysis.

A Lacanian Neuropsychoanalysis: Consciousness Enjoying Uncertainty builds on, among other sources of inspiration, Dall'Aglio's extremely fruitful conversations over recent years with the founding figure of (non-Lacanian) neuro-psychoanalysis, namely, Mark Solms. Dall'Aglio and Solms already publicly exchanged ideas with each other in the pages of the *Journal of the American Psychoanalytic Association* in 2021. Moreover, that same year, Solms published what is perhaps his most sophisticated and ambitious theoretical treatise to date: *The Hidden Spring: A Journey to the Source of Consciousness*. In this 2021 book, Solms weaves together neuroscience, psychoanalysis, and philosophy, relying especially broadly and deeply on the neuroscientific work of Jaak Panksepp (with his investigations into mammalian emotional systems) and Karl Friston (with his utilizations of the so-called "free energy principle," in which this "energy" is conceived of not as some sort of ethereal, fluid-like substance, but, instead, in strictly information-theoretic terms). Indeed, one way to read Dall'Aglio's *A Lacanian Neuropsychoanalysis* is as putting Lacan's theoretical and clinical ideas into mutually transformative dialogue with Solms's psychoanalytic and philosophical deployments of both Panksepp's affective neuroscience as well as Friston's models of predictions and prediction errors in the cerebral apparatus (with such predictions aiming to decrease "free energy," as surprising information, within the central nervous system).

To date, much of the limited Lacanian attention that has been paid to the neurosciences has tended to focus on the epigenetic and neuroplastic features of humans' brains. Coupled with Sigmund Freud's and Lacan's shared emphases on the importance of the biological fact of uniquely human prolonged prematurational helplessness for the ontogenetic formation of psychical subjectivity, a genetically-determined-not-to-be-genetically-determined human nervous system turns out to be evolutionarily pre-programmed to be culturally re-programmed; in other words, this system is naturally predisposed towards the dominance of exogenous nurture over endogenous nature. Helplessness inclines the young (proto-)subject specifically towards re-programming by the significant others upon which this vulnerable, impressionable being depends for its very existence and survival. In and through such more-than-natural, socio-linguistic mediations being brought to bear on malleable cerebral matter, the brain of neuroscience is, so to speak, transubstantiated into psyche of psychoanalysis.

However, this just-summarized linking of prematurational helplessness *à la* Freud and Lacan with epigenetics and neuroplasticity as more recent biological discoveries is only the beginning of what can and should be done in terms of developing a Lacanian neuro-psychoanalysis. Indeed, Dall'Aglio's book amply demonstrates just how much further one can go down this promising path. While incorporating the findings of prior efforts by others to interface Lacanianism with neurobiology, Dall'Aglio proceeds to explore sizable swathes of previously uncharted territory in terms of sites of consequential overlap between Lacanian psychoanalysis and the neurosciences (as well as the latter's uptake by non-Lacanian versions of neuro-psychoanalysis). So as to indicate better the originality of Dall'Aglio's contributions here, I briefly will turn to his deft neuro-psychoanalytic employment of Lacan's conceptualization of so-called "fundamental fantasies."

For Lacan as for Freud, the unconscious is the privileged and distinctive concern of both the theory and practice of psychoanalysis. Furthermore, in terms of Lacan's "formations of the unconscious" (as an unconscious formed *qua* "structured like a language"), the fundamental fantasy is a sort of master template pivotal both to a metapsychological rendition of subjectivity as well as to a clinical analytic treatment of symptoms and suffering. The Lacanian fundamental fantasy is symbolized by Lacan, in the formal language of his "mathemes," as $ ◊ a. What all of this means is that the fundamental fantasy is a sort of transcendental schema or matrix configuring a basic pattern of relation (◊ as condensing the symbols for conjunction [∧], disjunction [∨], less than [<], and greater than [>]) between a psychical subjectivity riven by tensions and antagonisms ($) with this subjectivity's libidinally charged objects and others (*a*). This unconscious schematism shapes and steers the desiring life of the speaking subject. As such, what Lacan calls the fundamental fantasy is crucial not only to a psychoanalytic account of who and what we are and how we live our lives, but also to an analytic alleviation of psychical pain and dissatisfaction resulting from persons' unwitting repetitious re-instantiations of the coordinates of these fantasies.

Dall'Aglio neuro-psychoanalytically recasts Lacan's fundamental fantasy, itself situated at the very heart of the unconscious as per Lacan, starting from a Lacanian reconsideration of the Pankseppian affective neuroscience upon which Solms's version of neuro-psychoanalysis heavily relies (in turn, Dall'Aglio persuasively problematizes a number of Lacanian orthodoxies regarding affects). Specifically, Panksepp identifies seven primary emotional circuits at the subcortical basis of all mammalian brains, human brains included. Solms adopts this taxonomy and makes it foundational for his own neuro-psychoanalytic depiction of humans' mental lives (with both Solms and Dall'Aglio identifying Panksepp's basic emotional circuits as neurobiological instances of Freud's drives [*Triebe*]).

In the spirit of the Lacan of *Encore* quoted and parsed by me at the start of the present foreword, one of Dall'Aglio's key moves in this context is to depict Panksepp's (and Solms's) set of seven elementary mammalian affective systems as a conflict-ridden natural Real akin to Lacan's symbolic order as a "barred" big Other, an internally and insurmountably inconsistent network of information-bearing components converging and clashing with each other. The nature that produces mammals' ensemble of primary emotions via evolutionary and genetic processes abandons humans especially to the abyss of their own uncertainties as to which emotional systems to prioritize over others and how to do so in the guise of action-guiding predictions satisfactorily responding to life's myriad unpredictable vicissitudes and challenges. In the absence of any (natural or otherwise) predetermined systematization of the neuro-psychical bases of mental life that would automatically adjudicate between basic affective circuits and dictate specific predictions governing behavior, the human being is compelled to become a subject divided ($) between its incompatibility-plagued, unsystematized plurality of primitive emotional systems and the impossible-to-master complexities of countless material and social externalities confronting it.

The inevitable errors of $'s predictions generate the surprises of excess free energy (again as per Friston's version of the free energy principle as taken over by Solms). This excess marks the failure fully to eliminate the "free energy" of neural excitation. And, as Dall'Aglio argues, such errors end up placing the barred subject of a disharmonious libidinal-affective economy in relation to a strange enjoyment (*jouissance*) of repeatedly producing and responding (in the guise of Freudian repetition compulsion [*Wiederholungszwang*]) to the excess free energy associated with specific failed predictions linked to the structural code of the fundamental fantasy. In light of Dall'Aglio's justifiable association of these compulsively repeated (and weirdly enjoyed) prediction errors with Lacan's *objet petit a*, one now can see why and how Dall'Aglio maintains that the Lacanian fundamental fantasy as $ \lozenge a$ can and ought to be brought into relationship with neurobiology via a suitably careful and sophisticated neuro-psychoanalysis.

As Dall'Aglio rightly suggests, fundamental fantasies and other formations of the unconscious responsible for symptomatic repetitions (with the latter encompassing a whole class of phenomena at the very core of the clinical analytic experience) are instances of "prematurely autonomized prediction." For Friston, Solms, and Dall'Aglio alike, the central nervous system tries as quickly and completely as possible to reduce its free energy to zero (echoing aspects of the later Freud's death drive [*Todestrieb*]) through successful signifier-like cognitive-ideational-representational predictions of what will quiet the energetic clamoring of its affective and motivational (i.e., libidinal) machinery. But, quickness is at odds with completeness in that the brain's prematurely autonomized predictions (in such guises as fundamental fantasies), given their inevitable prematurity, are prone to significant error and, hence, to the generation of the surprisingly addictive secondary gains of excess free energy as produced in and through specific prediction errors (i.e., the *jouissance* bound up with instantiations of *objet a*).

Although the brain is desirous of ridding itself of free energy, it is lazy about how it sets about doing so. Its autonomized predictions invariably are premature in one or more manners. The brain's laziness results in repeated failures that tend to perpetuate themselves, driving at least a few human organisms into such places as analysts' consulting rooms.

In *The Hidden Spring*, Solms thoughtfully engages with "the hard problem of consciousness" as canonically posed by Anglo-American Analytic philosopher of mind David Chalmers. Without diving into the details of this engagement on the present occasion, Chalmers's framing of his (in)famous "hard problem" brings into play, as a now well-known thought experiment, the figure of the philosophical zombie as a being physically and behaviorally identical with a living human but devoid of any consciousness whatsoever (i.e., a human-like automaton of sorts). Drawing on both Freud and Friston, Solms can be construed as insinuating that the figure of Chalmers's zombie is far from being merely the intellectual hypothetical of an armchair philosophical thought experiment. Instead, this figure represents the ultimate death-drive-type *telos* of the human central nervous system in general. Zombie-hood would be the brain's fantasy of itself as fully automated and, thus, freed from the burden of qualitative, affective consciousness.

In fact, thanks to Dall'Aglio's contributions in this same neuropsychoanalytic vein, one now can appreciate that the overall framework hierarchically organizing and prioritizing various affective circuits (as drives) and their typical predictions—this framework is nothing other than Lacan's fundamental fantasy—is the ultimate symptom-generating result of the lazy, self-zombifying human brain. Hence, Dall'Aglio additionally paves the way for importing the death drive, controversial both within and beyond analytic circles, into the neurosciences and neuropsychoanalysis. When introducing the *Todestrieb* in 1920's *Beyond the Pleasure Principle*, Freud brought his then-new metapsychological notion into connection with biological references, all the while acutely aware of the highly speculative nature of these connections due to the relatively underdeveloped state of the life sciences at the time. Seizing on the plethora of opportunities afforded by a century's worth of biological (especially neurobiological) advances since Freud's own era, Dall'Aglio convincingly demonstrates that even the most conceptually abstract and seemingly unnaturalizable components of Freudian and Lacanian metapsychologies are amenable to interfacing with sciences that themselves have become ever-more ripe and ready for being cross-fertilized with psychoanalysis.

Eric Kandel famously has insisted, including in his 2000 acceptance speech for the Nobel Prize in Physiology or Medicine acknowledging his discoveries regarding the biology of memory, that a thorough and non-reductive rapprochement between psychoanalysis and neurobiology will be central to twenty-first-century progress in illuminating human mindedness.[5] I confidently predict that *A Lacanian Neuropsychoanalysis* will succeed at making Lacan in particular indispensable for such a

---

[5] Eric R. Kandel, "The Molecular Biology of Memory Storage: A Dialogue between Genes and Synapses," Nobel Lecture, December 8, 2000, http://www.nobelprize.org/nobel_prizes/medicine/laureates/2000/kandel-lecture.pdf. Eric R. Kandel, "A New Intellectual Framework for Psychiatry," *Psychiatry, Psychoanalysis, and the New Biology of Mind*, Washington, D.C.: American Psychiatric Publishing, 2005, pg. 38). (Eric R. Kandel, "Biology and the Future of Psychoanalysis: A New Intellectual Framework for Psychiatry Revisited," *Psychiatry, Psychoanalysis, and the New Biology of Mind*, pg. 64.

rapprochement in the years and decades to come—perhaps to the surprise of many psychoanalysts and neuroscientists. Along related lines, Lacan, in his 1965 summary of his 1964 eleventh seminar on *The Four Fundamental Concepts of Psychoanalysis*, recommends moving from the more familiar, standard question "*Is psychoanalysis a science?*" to asking instead "*What would be a science that includes psychoanalysis?*"[6] I can think of no better answer to the latter question than Dall'Aglio's admirable book.

Albuquerque, New Mexico   Adrian Johnston
May 2024

---

[6] Jacques Lacan, "*Les quatres concepts fondamentaux de la psychanalyse: Compte rendu du Séminaire 1964*," *Autres écrits* [ed. Jacques-Alain Miller], Paris: Éditions du Seuil, 2001, pg. 187.

Acknowledgements

I first outlined these ideas while studying with Joan Copjec at Brown University. I am forever indebted to her and Azeen Khan for teaching me how to read the mazes of Lacan's speech. I wrote a trilogy of papers (with the encouragement of Mitchell Wilson) titled "Sex and Prediction Error" which was published in the *Journal of the American Psychoanalytic Association*. With the support of Derek Hook and Calum Neill, the editors of the Palgrave Lacan series, I expanded those ideas into the current book. To all of them, I owe great thanks.

Competing Interests The author has no conflicts of interest to declare that are relevant to the content of this book.

Ethics Approval All case material reported in this book has been disguised and deidentified.

Praise for *A Lacanian Neuropsychoanalysis*

"This book moves the needle moderating the debates between psychoanalysis and neuroscience. Rigorously schooled in the theories informing both sides, Dall'Aglio pokes and prods each side with the instruments and ideas forwarded by the other. Whether the two will ever move together as one is deeply uncertain, but the terms of their engagement will be altered by the book's impressive interventions."

—Joan Copjec, *Professor of Modern Culture and Media,*
Brown University, Providence, USA

"This is a masterful and scholarly account of Lacanian psychoanalysis—drilling down on the connection between affective consciousness and jouissance. The foundational (if elusive) Lacanian notion of jouissance is evinced with a clever triangulation using (i) Panksepp's affective neuroscience, (ii) Solms's neuropsychoanalysis and (iii) the theoretical neurobiology afforded by the free energy principle. With this conceptual triumvirate at hand, there is surely a fresh and revealing perspective on offer for every reader of this carefully crafted account of our psychic life. I thought the book's dénouement was an apt description of what it has to offer: 'although we do not know what will follow, that uncertainty becomes precise and precious.' In short, this book lays bare the angst and joys of resolving uncertainty—a surplus of prediction error—about ourselves and the Other."

—Karl J. Friston, MBBS, MA, MRCPsych, MAE, FMedSci, FRBS, FRS,
Scientific Director: Wellcome Centre for Human Neuroimaging,
UK; Professor: Queen Square Institute of Neurology, University |
College London, UK; Honorary Consultant: The National
Hospital for Neurology and Neurosurgery, UK

"This original and challenging book aims to create a dialogue between neuroscience and psychoanalysis which many have sought actively to avoid. Thought-provoking and stimulating, it will be an important reference for work at the intersection of these fields, and also for all students of Lacanian psychoanalysis."

—Darian Leader, *Psychoanalyst and member of the Centre*
for Freudian Analysis and Research, UK

"Writing a clear text on Lacan's approach to psychoanalysis is difficult enough. Like Scylla and Charybdis, the danger is found in failing to steer between the cliff of formalism on the one side and the rocks of obscurantism on the other. John Dall'Aglio succeeds in this perilous adventure and does so in the most challenging manner: by mapping Lacanian concepts onto contemporary neuroscience. Study of the interplay between psychoanalysis and neuroscience and our understanding of Lacanian theory are both profoundly enriched by the results of his journey.

Dall'Aglio set the terms for this challenge as a three-fold maneuver. This entails, first, to map Lacanian concepts in what he calls 'the neuroscientific space', then to observe their operation in that space, and finally, to return to the 'psychoanalytic space' to consider the effects. It is a clear and productive method and helps both the author and the reader to keep their bearings in this confusing sea.

Dall'Aglio's review of the objections to considering the interplay between psychoanalysis and neuroscience is thorough and fair. He is neither dismissive of the objections nor does he oversimplify them for the benefit of his own project. This review itself is an important and beneficial feature of the book.

There are two philosophical frameworks that support Dall'Aglio's project. One is that of Dual-Aspect Monism. As old as Spinoza and as current as Mark Solms, this framework allows the consideration of two realms of thought in a non-reductive interplay. The second is Transcendental Materialism as developed by Adrian Johnston which likewise meaningfully relates two registers that are viewed as essentially disjoint. Dall'Aglio combines these two frameworks in an original and effective manner.

The outcome of Dall'Aglio's voyage is an insightful and valuable study that is fair and respectful of the complexity and nuances of Lacanian thought as well as contemporary neuroscience and neuropsychoanalysis. Dall'Aglio's mapping in the space of neuroscience is knowledgeable, well-researched, and clearly written. And when he returns to the psychoanalytic realm, the benefits of the foray into neuroscience are clear and useful for the psychoanalytic reader.

Dall'Aglio's writing is clear and accessible which is a considerable achievement given the complexity of the two domains he is studying. This text should be of interest to readers of psychoanalysis at all levels of experience."

—David Lichtenstein, *NYU Postdoctoral Program in Psychoanalysis, New York, USA; Pulsion International Institute of Psychoanalysis, New York, USA; Institute for Psychoanalytic Training and Research, New York, USA; Psychoanalytic Institute of Northern California, San Francisco, USA*

"Dall'Aglio really understands neuropsychoanalysis, and he presents it lucidly. In addition, he contributes to it, in novel and important respects. By 'translating' Lacan into neuropsychoanalytic terms, he also enables one to understand him in a new way (and in my case, for the first time)."
—Mark Solms, *Director of Neuropsychology, Neuroscience Institute, University of Cape Town, South Africa*

"Lacan's brain would be a brain that enjoys itself far too much. An organ of thinking that is full of aberrations, loopholes and repetitions, whose evolution is disjunct and unpredictable, whose nature is fundamentally weak and porous. We don't know what will come of this brain whose emotional system is dysregulated by unchained affects with a sexual system that doesn't align with any idea of cortical processing. While many might think neuroscience to be anathema to Lacanian psychoanalysis, John Dall'Aglio shows us the value of this confrontation for revising our ideas about what it means to be hard-wired."
—Jamieson Webster, *Psychoanalyst and author of Disorganization and Sex; The New School for Social Research, New York, USA*

"John Dall'Aglio is a true treasure, a gift to psychoanalysis and to the burgeoning field of neuropsychoanalysis. In *A Lacanian Neuropsychoanalysis: Consciousness Enjoying Uncertainty*, Dall'Aglio forges completely original and compelling intersections and integrations among three seemingly distinct domains: Lacanian psychoanalysis, computational neuroscience, and the work of Mark Solms. The red thread that conceptually connects these three areas is sexuality and its attendant excesses, and the result of Dall'Aglio's effort is inspiring in the best sense. Anyone who opens this book and begins to encounter its treasures will quickly see that this is not hyperbole. One can only hope that *Lacanian Neuropsychoanalysis* is the first of many books to be penned by this most original thinker."
—Mitchell Wilson, *Editor-in-Chief Emeritus, Journal of the American Psychoanalytic Association, Training and Supervising Analyst, San Francisco Center for Psychoanalysis, San Francisco, USA; Training and Supervising Analyst, Psychoanalytic Institute of Northern California, San Francisco, USA*

"This book is like several books in one. Dall'Aglio offers a smart discussion of Lacan's essential concepts, provides a clear overview of the significance of neuropsychoanalysis, and, above all, fosters a dialectical encounter in the tradition of

Lacan himself, bringing together diverse scholarly perspectives that may not seamlessly align, prompting readers to continue reflecting."
—Stijn Vanheule, *Professor and Chair, Department of Psychoanalysis and Clinical Consulting, Ghent University, Belgium*

"The tension between brain sciences and psychoanalysis appears unsurmountable: there seems to be no common language between the two. Here enters Dall'Aglio: instead of accepting the gap or privileging one side as the truth of the other (like 'neuronal sciences make psychoanalysis obsolete'), he effectively does the impossible: based on a deep knowledge of neuronal sciences and of Lacanian psychoanalysis, he triumphantly succeeds in mediating the two, discovering libidinal mechanisms formulated by Lacan in the very core of how neuronal sciences describe the brain. Dall'Aglio's achievement is nothing less than epochal: nothing will be the same after *Lacanian Neuropsychoanalysis*, neither in neuronal sciences nor in psychoanalysis."
—Slavoj Žižek, *International Director, Birkbeck Institute for the Humanities, University of London; Senior Researcher, Department of Philosophy, University of Ljubljana*

Contents

Part I Can There Be a Lacanian Neuropsychoanalysis? 1

1 Introduction 3

2 Controversies, Criticisms, and Challenges of a Lacanian Neuropsychoanalysis 11

3 A Philosophical Basis for a Lacanian Neuropsychoanalysis 21

Part II The Enjoying Brain 41

4 The Concept of *Jouissance* 43

5 The Free Energy Principle 65

6 Mark Solms's Neuropsychoanalytic Meta-Neuropsychology 77

7 *Jouissance* is Surplus Prediction Error 97

8 The Neuronal Real: Antagonism Immanent to the Brain 111

9	Real, Imaginary, and Symbolic Knottings in the Predictive Model	123

Part III		Developing Implications of a Lacanian Neuropsychoanalysis	151
10		The Critique of *Jouissance*	153
11		A Neuropsychoanalytic Contribution to Debates over *Jouissance*	161
12		Affects like Signifiers	171
13		Toward Levels of the Symbolic	189
14		Clinical Lacanian Neuropsychoanalysis	201
15		Conclusion	239

List of Tables

Table 6.1 Panksepp's seven basic emotional systems 81
Table 8.1 Lacanian concepts and their proposed neuropsychoanalytic correlates 120
Table 9.1 Lacanian concepts and their proposed neuropsychoanalytic correlates 146

Part I

Can There Be a Lacanian Neuropsychoanalysis?

1

Introduction

Abstract My first encounter with neuropsychoanalysis was a video of Mark Solms's lecture "The conscious id." I thought to myself: "Wait, I learned that the id is unconscious. This can't be right." So, I watched the video, and I have been fascinated by neuropsychoanalysis ever since. Perhaps with some personal enjoyment of impossibility, I have specifically tried to integrate the "impossible" Lacanian domain of sexuality and *jouissance* with contemporary neuropsychoanalysis, affective neuroscience, and computational neuroscience. This chapter introduces the trajectory of this book, my synthesis of a Lacanian neuropsychoanalysis.

Keywords Neuroscience • Lacan • Neuropsychoanalysis • Freud • Solms • Sex • Jouissance • Panksepp • Free energy principle • Friston

As a high school senior, I was first introduced to psychoanalytic theory through Charles Brenner's (1974) *An Elementary Textbook of Psychoanalysis*. We read two chapters on Freud's structural model. I clearly recall my excited curiosity when learning about conflict between different psychical agencies: the naughty, sexual, aggressive id; the nasty, sadistic superego; and the poor little ego caught in the middle. We also learned some introductory brain functions. When we learned about conflict between a "rational" prefrontal cortex response and an "irrational" amygdala response, my ears perked. Was there any similarity between these two theories I'd learned?

I Googled "neuroscience psychoanalysis" and discovered the field of neuropsychoanalysis. Fascinated, I read Mark Solms's and Oliver Turnbull's (2002) *The Brain and the Inner World* and prepared to give a presentation on the material at the end of the year. One friend had been critical of my "Freud-kick" at the time. He constantly teased me about how Freud made everything about sex, Freud just had his own sexual problems, and so on. In hindsight, this probably got under my skin, as I vehemently refused and debated his criticism of Freud as sexually-reductive. With Solms's use of affective neuroscience, things were not just about sex: there was also attachment, fear, nurturing care, curiosity, rage, and so on.

After I gave the presentation on neuropsychoanalysis, there was time for questions and discussion. I jumped around the PowerPoint slides as we shifted from topic to topic. With a cheeky look on his face, this friend raised his hand to ask a question. In this more formal setting, he dared to open this can of worms (and what sort of psychoanalytic symbol could *worms* possibly be, or a *can* for that matter?) and asked something like *Don't you think Freud makes everything about sex*? I dug my heels in the ground, gesticulated my hands around my head as I re-iterated Jaak Panksepp's seven emotional drives, and spurt out something about conflict and pleasure in an expanded, general sense. Standing in front of the PowerPoint screen, I concluded the act and exclaimed: "Guys, psychoanalysis isn't about sex!" And behind me, in towering font on the slides, were enormous boxes saying: "Freud's Psychosexual Stages: Oral, Anal, Phallic, Latency, Genital."

This book is, in a certain sense, an attempt to situate and centralize the radicality of sexuality in neuropsychoanalysis. Lacan's term *jouissance* seeks to capture the perversity, excessiveness, and paradoxical blending of pleasure and unpleasure in Freud's notion of sexuality. These dimensions easily drop out, sometimes within Freud's own writings and often when trying to bridge Freud to other disciplines like neuroscience.

For instance, Jaak Panksepp (1998) identified seven emotional instincts across all mammals, one of which is LUST.[1] He primarily studied non-human mammals using deep brain stimulation and pharmacological methods. In 2003, Kenneth Davis, Jaak Panksepp, and Larry Normansell created the Affective Neuroscience Personality Scales (ANPS) to measure these systems in humans via a Likert-scale self-report questionnaire. However, they only included *six* systems in their questionnaire. The one excluded? LUST! While there was rational(e) for not including questions about sexual experiences (concern over response bias; the lack of a "sex factor" in the Big Five model; Montag et al., 2021), the psychoanalytic reader cannot help but wonder if some repression of sexuality also took place. Nearly two decades passed after the publication of the original ANPS before a LUST scale was developed (Fuchshuber et al., 2022).

But even questions about sexual activity like those asked in Fuchshuber et al.'s LUST scale do not capture the polymorphous perverse nature of the sexual drive as described by Freud. Libido is not restricted to sexuality in the "veterinary sense," to borrow a phrase that Solms uses to describe Panksepp's LUST instinct. Lacan describes the paradoxical potential for sexual enjoyment to shift beyond the genitals:

> Sublimation is nonetheless satisfaction of the drive, without repression. In other words, for the moment, I am not fucking. I am talking to you. Well! I can have exactly the same satisfaction as if I were fucking. (Lacan, 1964, pp. 165–166)

Here, then, is the puzzle I had myself repressed when giving that presentation—with the ironic and immediate "return of the repressed" in the PowerPoint slides. Psychoanalysis is not (only) about *veterinary* sexuality and genital intercourse. But it certainly is about sex. However, what is this strange sexuality that psychoanalysis deals with? And could we think about this strange sexuality—this strange enjoyment that Lacan calls *jouissance*—in the brain?

---

[1] Capitalization of these systems follows Panksepp's lexicon to designate specific neural circuits, beyond semantic associations.

In 2015, Mark Solms published a collection of essential papers on neuropsychoanalysis titled *The Feeling Brain*. One of Solms's major contributions is the integration of Jaak Panksepp's affective neuroscience into psychoanalysis (Solms, 2013). This reverberates in Solms's focus on affect as the most basic form of consciousness and his thesis of the conscious id. Multiple revisions to psychoanalytic theory follow from this paradigm shift, including revisions to drive theory, the Oedipus complex, and clinical formulation.

The ideas for this book arose from a Lacanian perspective on this element of Solms's work. Where non-Lacanian psychoanalysis and neuroscience speak of affects, Lacanian psychoanalysis speaks of *jouissance*. What are the differences between affect and *jouissance*? Can *jouissance* be situated in the brain? What if the brain does not simply feel? What if the brain *enjoys*? Could there be a neuropsychoanalysis of *The Enjoying Brain*?

As one might expect, Lacanian neuropsychoanalysis is not without controversy. In Chap. 2, I describe the principal criticisms of dialogue between Lacan and neuroscience—criticisms which some suspicious readers might share. As I discuss in Chap. 3 and develop throughout the book, I hope to assuage any concerns over bio-reductionism or ethico-clinical normalization.

To some readers, this book might be somewhat dense. Admittedly, it integrates concepts from three different fields: Lacanian psychoanalysis, computational neuroscience, and Solmsian neuropsychoanalysis. Each is complex in its own right. I lay out the foundational concepts in these fields in Chap. 4 (the Lacanian concept of *jouissance* in relation to the real, symbolic, and imaginary registers), Chap. 5 (Karl Friston's Free Energy Principle, the Bayesian brain, and predictive coding), and Chap. 6 (Solms's neuropsychoanalytic incorporation of the Free Energy Principle and Jaak Panksepp's affective neuroscience). These chapters set the backdrop for my argument that *jouissance* finds its correlate in the computational neuroscientific concept of *surplus prediction error* (Chap. 7). Correspondingly, I also argue that predictions can be understood as signifiers.

Chapter 8 details specific elements of Solms's theory (often overlooked) that demonstrate the centrality of structural antagonism for affective consciousness. This allows me to draw out a precise difference between (and entwining of) *jouissance* and Panksepp's emotional systems. *Jouissance*, as the surplus prediction error of affective consciousness, operates in the empty space of the barred subject, divided in the antagonism of innate emotional systems.

In Lacanian register theory, there are not simply real, imaginary, and symbolic dimensions. These registers are *knotted* together, intertwined, and conjoined at crucial points of subjective structure. Taking this idea as a conceptual tool (Verhaeghe, 1999), I assert that these registers are knotted in the brain. Specifically, I draw upon the Lacanian fundamental fantasy to conceptualize real-symbolic-imaginary knotting in the brain. Chapter 9 details certain knottings and a specification of *objet a* as a residual of uncertainty in the predictive field of the Other.

In Chaps. 10 and 11, I demonstrate how Lacanian neuropsychoanalysis can contribute to debates in Lacanian literature, namely criticisms of the concept of *jouissance* from Darian Leader. Dialogue between neuroscience and Lacan opens new ideas regarding *jouissance* as well as the organization of Panksepp's basic emotions. I then develop consequences of this model, namely that affects operate like signifiers (Chap. 12) and that one can discern levels of the symbolic separate from language *per se* in the brain (Chap. 13).

In Chap. 14, I specify a clinical Lacanian neuropsychoanalytic perspective: how Lacanian psychoanalysis can enrich clinical neuropsychoanalysis and vice versa. This includes a conceptualization of punctuation and scansion in neuropsychoanalytic terms. Moreover, I put forward a model of Lacanian clinical formulation focused on the metabolization of prediction error and the modulation of automatized predictions. My hope is that clinicians, researchers, and theorists will find that these ideas spark creative uncertainty that advances interdisciplinary work and dialogue.

Before we begin, a note on terminology is in order. Throughout the book, especially when integrating Lacan with neuroscience, I employ (and create) Lacanian algebraic symbols or what he called "mathemes" (e.g., $, *a*, S1, S2, J) to guide this theoretical development. I do so for the same reason that Solms (2020) retains the Greek letters of Freud's *Project* and employs the mathematical symbols of Friston's Free Energy Principle. Abstraction assists in detaching from phenomenological description, to conceptualize features of the laws governing the mental apparatus (Solms, 2021). Of course, my Lacanian algebraic symbols are quasi-mathematical and purely serve the goal of abstraction.

Concerning these symbols (and Lacanian concepts more broadly), I share Solms's sentiment on his rewriting of Freud's *Project* through contemporary neuroscience:

> the Greek letters used here are purely conventional; they have no literal meaning. Also, although the symbols denote concepts that are equivalent to those that Freud used, they are not identical with them; this is because the concepts have been substantively updated. (Solms, 2020, p. 32, fn. 2)

My Lacanian (neuropsychoanalytic) terms and symbols are equivalent to those used by Lacan but not identical, for they have been "substantively updated." Incorporation of these symbols serves a *thinking with concepts* that seeks to "continue the project that…[Lacan] had opened" (Leader, 2021, p. 134). Besides, in good Lacanian fashion, meaning does not remain fixed!

References

Brenner, C. (1974). *An elementary textbook of psychoanalysis*. Anchor Books.
Davis, K., Panksepp, J., & Normansell, L. (2003). The affective neuroscience personlaity scales: Normative data and implications. *Neuropsychoanalysis,* 5(1), 57–69. https://doi.org/10.1080/15294145.2003.10773410
Fuchshuber, J., Jauk, E., Hiebler-Ragger, M., & Unterrainer, H. (2022). The affective neuroscience of sexuality: Development of a LUST scale. *Frontiers in Human Neuroscience, 16,* 853706. https://doi.org/10.3389/fnhum.2022.853706
Lacan, J. (1964). *The seminar of Jacques Lacan, Book XI: The four fundamental concepts of psychoanalysis* (J.-A. Miller, Ed., A. Sheridan, Trans.). Norton.
Leader, D. (2021). *Jouissance: Sexuality, suffering and satisfaction*. Polity.
Montag, C., Elhai, J., & Davis, K. (2021). A comprehensive review of studies using the affective neuroscience personality scales in the psychological and psychiatric sciences. *Neuroscience & Biobehavioral Reviews, 125,* 160–167. https://doi.org/j.neubiorev.2021.02.019

Panksepp, J. (1998). *Affective neuroscience: The foundations of human and animal emotions*. Oxford University Press.

Solms, M. (2013). The conscious id. *Neuropsychoanalysis, 15*(1), 5–19. https://doi.org/10.1080/15294145.2013.10773711

Solms, M. (2015). *The feeling brain: Selected papers on neuropsychoanalysis*. Routledge.

Solms, M. (2020). New project for a scientific psychology: General scheme. *Neuropsychoanalysis, 22*(1–2), 5–35. https://doi.org/10.1080/15294145.2020.1833361

Solms, M. (2021). *The hidden spring: A journey to the source of consciousness*. Profile Books.

Solms, M., & Turnbull, O. (2002). *The brain and the inner world: An introduction to the neuroscience of subjective experience*. Other Press.

Verhaeghe, P. (1999). *Does the Woman exist? From Freud's hysteric to Lacan's feminine* (M. du Ry, Trans.). Other Press.

2

Controversies, Criticisms, and Challenges of a Lacanian Neuropsychoanalysis

Abstract Lacanian psychoanalysis is largely, and notoriously, antinaturalist. There are special challenges in attempting to formulate a specifically *Lacanian* neuropsychoanalysis. Many Lacanians have criticized neuropsychoanalysis for bio-reductionism, ethical normalization, and an inability to capture the formal contradictions of subjectivity articulated in Lacan's meta-psychology. This chapter summarizes the arguments behind these criticisms, to contextualize this book in the larger field of controversies over (Lacanian) neuropsychoanalysis. These criticisms will be addressed over the course of this book.

Keywords Reductionism • Ethics • Real • Neuroscience • Lacan • Subject • Normative • Instinct • Drive

---

This chapter summarizes and expands arguments from Dall'Aglio (2020).

© The Author(s), under exclusive license to Springer Nature Switzerland AG 2024
J. Dall'Aglio, *A Lacanian Neuropsychoanalysis*, The Palgrave Lacan Series,
https://doi.org/10.1007/978-3-031-68831-7_2

In the summer of 2019, the International Neuropsychoanalysis Association held its annual congress in Brussels, Belgium. "Sex, Drive, and Enjoyment" was the theme, and it was the first congress to explicitly call for Lacanian perspectives on these topics in neuropsychoanalysis. A week earlier, the World Association of Psychoanalysis (an international Lacanian psychoanalytic organization) held its annual congress in the same city. Its argument: "The Unconscious and the Brain: Nothing in Common." An amusing coincidence?

A "Lacanian neuropsychoanalysis" seems like an oxymoron. Lacanian psychoanalysis prioritizes the human subject as denaturalized, cut-off from biological determinism, marred by its immersion in language and culture. Lacan often took an anti-naturalist position, privileging abstraction, mathematics, linguistics, philosophy, and topology to conceptualize and situate psychoanalysis. On the other hand, few neuroscientific researchers have any familiarity with Lacan. Lacanian rejections of biology, phylogeny, and core natural scientific ideas like homeostasis are often anathema to neuroscientists. Developing a "Lacanian neuropsychoanalysis" may seem like an impossible endeavor that is unlikely to garner the attention of mainstream Lacanians or neuroscientists. But perhaps this is precisely why such a project has such potential to open unique perspectives in both disciplines. It walks an unexpected tightrope in a field of impossibility.

To begin this book on a Lacanian neuropsychoanalysis, I will review major criticisms levied by Lacanians against the neuropsychoanalytic project (Ferraro, 2022; Laurent, 2014; Vanderveken, 2018). These run as follows: (1) neuroscience is bio-reductive, (2) neuroscience is ethically normative, and (3) neuroscience cannot capture the real. Many of these criticisms rely on presumptive readings of neuroscience or ignore a distinction between neuroscience as a scientific practice and neuroscientific discourse. By neuroscientific discourse, I refer to discourses that take up neuroscientific ideas, rather than neuroscientific findings *per se* (Sandberg, 2019). I will also reference non-Lacanian criticisms of neuropsychoanalysis to situate the Lacanian critiques in a broader psychoanalytic context. This chapter sets up the challenges to a Lacanian neuropsychoanalysis that I intend to overcome over the course of this book.

Neuroscience Is Reductive

One major project within neuropsychoanalysis is the mapping of psychoanalytic concepts to the brain. Mark Solms's work is exemplary here, having charted neuro-functional localizations of the id and the ego (see Chap. 6; Solms, 2013). Such mappings have implications for theoretical revisions (Solms, 2021a, 2021b) and changes to clinical technique (Solms, 2018). Importantly, there are debates over Solms's model from within the neuropsychoanalytic field, with some preferring a psychological level of organization (e.g., primary and secondary processes; Bazan, 2023) or brain activity dynamics (e.g., spatiotemporal rhythms; Northoff & Scalabrini, 2021) over functional neuroanatomy. Nevertheless, a common principle in the field has been the effort to link psychoanalytic ideas (topographies, processes, dynamics, etc.) to the brain.

Critics of neuropsychoanalysis claim that neurofunctional mapping leads to treating the patient as a brain rather than a subject (Ferraro, 2022). Contemporary discourse indeed privileges the signifier "neuro-" for its supposed potential to enhance various fields: neuro-economics, neuro-education, neuro-humanities, and so on (Vanderveken, 2018). Invoking neuroscience often affords a weight or gravity that points to the contemporary sense that the brain is more real than the mind. Relying purely on brain-based understandings of human behavior risks glorifying the brain as "more real" than ephemeral phenomena like human subjects.

From a non-Lacanian standpoint, Blass and Carmeli (2007) critique neuropsychoanalysis for shifting the focus from meaning-making practice to irrelevant questions of brain correlates. They claim that neuroscience cannot *meaningfully* inform psychoanalysis. Psychoanalysis deals with psychological phenomena that are first discovered (and described) psychologically: drive, feeling, attachment, object relation, and so on. Neuroscience does not discover a drive; neuroscience identifies the neurobiological *correlates* of phenomena first discerned psychologically. While neuroscience can add knowledge of the material substrate to these phenomena, any knowledge of those phenomena first derived from the psychological sphere. Knowledge about these phenomena therefore flows from the psychological to the neurobiological, not vice versa. Therefore, neuroscience has nothing to add.

Blass and Carmeli (2007) claim that neuropsychoanalysis reduces the clinical experience to biological explanation—thereby shutting down the meaning-making process of an analysis. For example, a patient suffering from anxiety and fatigue spoke about feeling "stressed." They wondered how much of it is "nature versus nurture" and said "you might know since you have studied psychology." I could have provided an explanation of the hypothalamic-pituitary-adrenal axis (a brain-body neuroendocrine system involved in the stress response) and the role of both genetics and interpersonal experiences in the development of distress and bodily symptoms (Landa et al., 2012). This could be cast as addressing the patient as a brain (with a hypothalamic-pituitary-adrenal axis) rather than as a subject. Addressing the patient as a subject might instead attune to how the patient is making sense of their distress: split between something related to childhood experiences ("nurture") and something supposedly predetermined ("nature").

Of course, Lacanians are not concerned with the subject as meaning-making so much as desiring and enjoying (Fink, 2011). Nevertheless, these critics share common concerns. Ferraro (2022) suggests that neuropsychoanalysis would be more concerned with brain imaging data than social structures, for instance, to answer clinical questions. To return to my example above, a Lacanian might be interested in the network in which this symptom is situated: the address to an Other supposed to know (in the transference, I am one who knows something about their suffering because I study psychology), how their symptom allows them to sustain desire and procure enjoyment, and how this matrix fits within the backdrop of their history. Reference to neurological language risks obscuring these dimensions. Vanderveken's (2018) critique is precise: the brain "veils the real of jouissance." Something is lost in translation into neuro-cognition (Laurent, 2014). Although few reject that the brain is the material substrate for the mind, the neurobiological register is deemed epistemologically irrelevant to the play of signifiers and exchange of speech that constitutes the focus of Lacanian psychoanalysis (Ferraro, 2022). Dialogue with neuroscience is a bio-reductive move that risks replacing (and thereby losing) psychoanalytic concepts with an objective epistemology that forecloses the human subject.

Neuroscience Is Normative

Vanderveken further critiques the predominance of neuro-imaging: an *"ethics of desire is opposed to this civilization of the cipher* and of cerebral imagery" (Vanderveken, 2018, emphasis in the original). The "civilization of the cipher" *through* "cerebral imagery" fits with a psychotherapeutic tendency toward normalization. Pathology is redefined as deviation from an average, visualized through comparisons of brain activity, with patterns deemed statistically above or below the mean. The image of the brain brings a trojan horse of normalizing ethics that advocates for a return to the happy medium, whether by an expensive pill cocktail or by "correction" of unwanted thoughts or actions (Ferraro, 2022).

Moreover, this cerebral imagery "civilizes" the "cipher." For Lacan, formations of the unconscious are ciphered. Dreams, slips, and symptoms are disguised through the play of the signifying chain (condensation, substitution, displacement, and so on). Interpretation is (to some degree) a deciphering (Fink, 2011). Importantly, there is no universal "key" to this process—like a standardized codebook of dream-symbols. The signifiers involved in (de)ciphering are specific to each subject's speech and history.

One might understand the "civilization of the cipher" through the lens of Blass and Carmeli's (2007) critique. Rather than opening the specific speech (or meaning-making process) of the individual patient, the neuroscientific lens implies a translation from singular to universal (Last, 2021). Anxiety becomes excessive activation of the amygdala. Difficulty managing affects becomes a problem of prefrontal cortical capacities. Such neurobiological references make no note of individual history where idiosyncratic deciphering might occur. The brain appears as a universal codebook that ciphers the symptom (i.e., re-disguises it in neural terms) while also civilizing it (i.e., normalizing its cause, now in the brain, and its solution—pharmaceuticals, moving to an average, etc.).

Opposed to a cerebral normalization of the cipher, Vanderveken places an "ethics of desire." Psychoanalytic praxis ethically orients toward the patient as a desiring subject (Israely, 2018). The subject emerges in the formations of the unconscious as an effect of speech. In Lacanian algebra, the subject is written as \$—S-barred—the divided subject split between signifiers that do not adequately represent the subject (Žižek, 2020a).

For example, a patient felt that they saved their friend's children by taking care of them when their friend was financially struggling: "I saved her kids." After describing the situation in more detail, she slipped: "She saved our kids." One can speculate on the multiple potential chains (and the story overall): identifying as a mother, a wish to be saved, a wish for her own kids to be taken off her hands, frustration against the friend, and so on. The slip does not reveal the truth of the subject *per se*—rather, the subject as desiring emerges in the *act* of the slip, the breakage from the pre-formed narrative that coincides with the emergence of the novel (Israely, 2018). The subject *emerges* as divided *between* these signifiers, rather than being revealed as "truly there" in the slip or in the latent contents of the dream (Žižek, 2020a).

Desire is oriented toward what is not captured or what is enigmatic in speech and social discourse at large (designated by the term *objet a*; see Chap. 4). It is metonymic and capricious, ever shifting among objects and signifiers that do not quite fit the bill (Lacan, 1964). Psychoanalysis is ethically oriented toward this dehiscence in speech to promote a space for desire to emerge (Israely, 2018). This stance opposes a cerebral (de)ciphering insofar as the orientation toward novelty and contradiction does not fit with the universalizing image of the brain.

For Ferraro (2022), this difference between universal and singular is an ethical issue. Not only does neuropsychoanalysis introduce a universal neural reference for singular subjective speech; the pivot to a normalizing stance obliterates the possibility for the subject to speak. The "ethics of desire" is not merely a theoretical point; for Lacan, the subject assumes its position because it speaks. There is an ethical obligation to speak (and be heard as speaking, listening to the letter of the patient's speech) because the subject comes into being through this act (Lacan, 1959–1960, 1964). Lacanian psychoanalysis works in the reverse direction: from the universal to the singular, homing on the singular mode in which one makes do with the trauma of the drive (see Chap. 14; Lacan, 1972–1973; Miller, 2023).

Neuroscience Does Not Capture the Real

In Lacanian theory, co-incidental with the subject as divided among signifiers ($) is some surplus or excess equally not captured by representation (*objet a*; see Chap. 4). There is a "bulge in the field of perception" (Soler, 2015) that is not captured in the fMRI scanner (Ferraro, 2022). The "real" is the term for this negative excess that cannot be imaged or grasped (see Chap. 4). The *image* of the brain "veils the real"; neuroscience cannot attend to the emergent excess uncaptured by representation (Vanderveken, 2018).

This concern is more unique to Lacanian critics of neuropsychoanalysis. The issue is not only the spread of "neuro-" master discourses. Neuroscience—the image of the brain, the formulas for its synaptic activity, its models for predicting human behavior—denies any notion of a structural lack or positive surplus that escapes symbolization (Ferraro, 2022). Neuroscience's denial of contradiction or impossibility renders it incapable of attending to the real.

In contrast, Lacanian psychoanalysis "treats the real" by means of the symbolic by privileging the materiality (as opposed to meaning) of speech and its cracks in self-reflective discourse (Lacan, 1964). It affects what is not captured by speech (i.e., the real) by attending to the very failures of speech (Soler, 2015). For example, a patient described feeling "tired" in her body when she felt anxious or angry. A Lacanian ear would be less concerned with this sense-making or the mechanisms of her emotional arousal and more interested in the resonance of the signifier "tired." Indeed, she described her father (from whom she claimed to be very different) as "tired" and spoke about her grandfather whose "*reti*re*ment*" heavily impacted the family structure and his own mental health. The signifier *tired* did not adequately describe her emotional experience (i.e., its imprecision between anxiety and anger) and instead indexed a disturbing excess to her experience (i.e., the unmanageable nature of affects) that was ciphered in her family history. The "real" names the crack within or failure of the symbolic network to reach clear comprehension, the symbolic's own structural destabilization (Žižek, 2020b). According to critics of neuropsychoanalysis, this real (as both contradiction and disturbing excess) cannot be captured by any dialogue with the neurosciences (Ferraro, 2022).

To sum up, Lacanian critics take epistemological and ethical issues with neuropsychoanalysis. The shift to neuroscience is a bio-reductive move that replaces the singularity of speech with reference to neurobiology. Whether through statistical quantification or fMRI, neuroscience obscures (1) the idiosyncratic subject that emerges as divided through the slippages of speech and (2) the unnamable excess that eludes the grasp of representation. In place of a clinical ear toward capricious desire and a singular mode of enjoyment, neuroscience ushers a normalizing approach that treats deviations from the norm as something to be corrected. As Lacan put it in 1974:

> First off, let's get rid of this average Joe, who does not exist. He is a statistical fiction. There are individuals, and that is all. When I hear people talking about the guy in the street, studies of public opinion, mass phenomena, and so on, I think of all the patients that I've seen on the couch in forty years of listening. None of them in any measure resembled the others, none of them had the same phobias and anxieties, the same way of talking, the same fear of not understanding. Who is the average Joe: me, you, my concierge, the president of the Republic? (Lacan, reported in Skinner, 2014)

References

Bazan, A. (2023). Primary and secondary process mentation: Two modes of acting and thinking from Freud to modern neurosciences. *Neuropsychoanalysis*. https://doi.org/10.1080/15294145.2023.2284697

Blass, R., & Carmeli, Z. (2007). The case against neuropsychoanalysis: On fallacies underlying psychoanalysis' latest scientific trend and its negative impact on psychoanalytic discourse. *International Journal of Psychoanalysis, 88*(1), 19–40. https://doi.org/10.1516/6NCA-A4MA-MFQ7-0JTJ

Dall'Aglio, J. (2020). No-Thing in common between the unconscious and the brain: On the (im)possibility of Lacanian neuropsychoanalysis. *Psychoanalysis Lacan, 4*. http://psychoanalysislacan.com/issue-4/

Ferraro, D. (2022). The problem with neuropsychoanalysis – A reply to John Dall'Aglio. *Psychoanalysis Lacan, 5.* https://lacancircle.com.au/psychoanalysislacan-journal/psychoanalysislacan-volume-5/the-problem-with-neuropsychoanalysis-a-reply-to-john-dallaglio/

Fink, B. (2011). *Fundamentals of psychoanalytic technique: A Lacanian approach for practitioners.* Norton.

Israely, Y. (2018). *Lacanian treatment: Psychoanalysis for clinicians.* Routledge.

Lacan, J. (1959–1960/1992). *The seminar of Jacques Lacan, Book VII: The ethics of psychoanalysis* (J.-A. Miller, Ed., & D. Porter, Trans.) Norton.

Lacan, J. (1964/1978). *The seminar of Jacques Lacan, Book XI: The four fundamental concepts of psychoanalysis* (J.-A. Miller, Ed., A. Sheridan, Trans.). Norton.

Lacan, J. (1972–1973/2000). *The seminar of Jacques Lacan, Book XX: On feminine sexuality, the limits of love and knowledge* (J.-A. Miller, Ed., B. Fink, Trans.). Norton.

Landa, A., Peterson, B., & Fallon, B. (2012). Somatoform pain: A developmental theory and translational research review. *Psychosomatic Medicine, 74*(7), 717–727. https://doi.org/10.1097/PSY.0b013e3182688e8b

Last, C. (2021). The difference between neuroscience and psychoanalysis: Irreducibility of absence to brain states. *Neuropsychoanalysis, 23*(1), 27–38. https://doi.org/10.1080/15294145.2021.1926312

Laurent, É. (2014). *Lost in cognition: Psychoanalysis and the cognitive sciences* (A. Price, Trans.). Karnac Books.

Miller, J.-A. (2023). *Analysis laid bare.* Libretto Press.

Northoff, G., & Scalabrini, A. (2021). "Project for a spatiotemporal neuroscience" – Brain and psyche share their topography and dynamic. *Frontiers in Psychology, 12,* 717402. https://doi.org/10.3389/fpsyg.2021.717402

Sandberg, L. (2019). Interpreting neuroscientific facts. *Psychoanalytic Inquiry, 39*(8), 596–606. https://doi.org/10.1080/07351690.2019.1671124

Skinner, J. (2014, July 22). 'There can be no crisis of psychoanalysis' Jacques Lacan Interviewed in 1974. *Verso.* https://www.versobooks.com/blogs/news/1668-there-can-be-no-crisis-of-psychoanalysis-jacques-lacan-interviewed-in-1974

Soler, C. (2015). *Lacanian affects: The function of affect in Lacan's work* (B. Fink, Trans.). Routledge.

Solms, M. (2013). The conscious id. *Neuropsychoanalysis, 15*(1), 5–19. https://doi.org/10.1080/15294145.2013.10773711

Solms, M. (2018). The neurobiological underpinnings of psychoanalytic theory and therapy. *Frontiers in Behavioral Neuroscience, 12*, 294. https://doi.org/ 10.3389/fnbeh.2018.00294

Solms, M. (2021a). A revision of Freud's theory of the biological origin of the Oedipus complex. *Psychoanalytic Quarterly, 90*(4), 555–581. https://doi.org/10.1080/00332828.2021.1984153

Solms, M. (2021b). Revision to drive theory. *Journal of the American Psychoanalytic Association, 69*(6), 1033–1091. https://doi.org/10.1177/000 3065121105

Vanderveken, Y. (2018, November 26). *The Argument*. PIPOL9: https://www.pipol9.eu/the-argument-pipol9/?lang=en

Žižek, S. (2020a). *Hegel in a wired brain*. Bloomsbury.

Žižek, S. (2020b). *Sex and the failed absolute*. Bloomsbury.

3

A Philosophical Basis for a Lacanian Neuropsychoanalysis

Abstract Neuropsychoanalysis places subjective and objective perspectives on equal epistemological footing. This philosophical position of "dual-aspect monism" sees mind and brain as two appearances of the same part of nature. Here, I propose a method for shifting between these perspectives, to build a non-bio-reductive meta-neuropsychology. Additionally, by drawing on Adrian Johnston's Transcendental Materialism, I develop a specific philosophical basis for a Lacanian neuropsychoanalysis which allows a formulation of the antagonistic real in nature itself. I also specify a Lacanian neuropsychoanalytic clinical position which explicitly aims at non-normative intervention.

Keywords Transcendental materialism • Dual-aspect monism • Reductionism • Homeostasis • Neuroscience • Lacan • Solms • Dynamic localization • Instinct • Drive

Here, I will put forward a philosophical framework for the basis of a Lacanian neuropsychoanalysis that avoids the criticisms discussed in the

---

This chapter expands arguments from Dall'Aglio (2022).

previous chapter. I begin by reviewing the philosophical foundations of traditional (non-Lacanian) neuropsychoanalysis. I then discuss a specifically Lacanian approach to this foundation through the framework of Transcendental Materialism (Johnston, 2019).

Throughout, several ideas will be referenced that are fully developed later in the book. This is because some of the principles find support from neuropsychoanalytic integrations that require a full development of elements from different fields (affective neuroscience, computational neuroscience, Freudo-Lacanian theory). The desiring reader who returns to these principles after completing Sections 2 and 3 may appreciate them in greater depth.

Epistemological Bridges: Solmsian Dual-Aspect Monism

What is more real, thunder or lightning?

I ask this question (inspired by Solms, 2020b) when introducing students to a neuropsychoanalytic way of thinking about the relationship between mind and brain. Most of the time, students (who are typically not studying physics) will waver but lean toward lightning because it can "strike" things, cause injury, because it is seen before thunder, and so on. I then reveal that it is a trick question (or a Lacanian forced choice; indeed, one sharp student took issue with my phrase "more real"). It makes no sense to ask whether lightning or thunder is more real. *In nature*, there is a potential difference of electrical charge between two points such that a particle jumps from one point to the other. In this rapid movement (and friction with air particles), energy is released. Visually, this energy is *seen* as lightning. Audially, this energy is *heard* as thunder. Lightning and thunder are two *phenomenological perspectives* of the same *thing* in nature.

Neuropsychoanalysis takes an analogous view of the mind-brain relationship. Within the framework of Spinozian dual-aspect monism, mind and brain are two (incomplete) perspectives of the same thing in nature (Solms & Turnbull, 2002). Freud called this thing the "mental apparatus." One can look at this thing objectively—as an *object*. A patient came

3 A Philosophical Basis for a Lacanian Neuropsychoanalysis   23

in with depression. Objectively—were one to conduct fMRI scanning and neurophysiological tests—one might find dysregulated dopamine and serotonin systems. This view looks at the mental apparatus as an object called the brain. You can scan it, poke it, inject it with drugs, and so on. This same patient also spoke about this depression: they felt hopeless, lethargic, and self-critical. This view looks at the mental apparatus subjectively, as a *subject* with thoughts, feelings, and the like. Such a viewpoint takes advantage of a unique property of this part of nature—namely, that we know what it is like to *be* this part of nature. The mental apparatus is a strange torsional point where nature twists and looks upon itself in self-reflective consciousness of its own subjective experience.

Mark Solms calls upon dual-aspect monism as the philosophical basis for neuropsychoanalysis. For Solmsian dual-aspect monism, mind and brain are two phenomenological perspectives of the same ontological thing. Subjective phenomena like feelings and thoughts are no less real than neurobiological phenomena such as neurotransmitters and anatomical structures. Solms's position resembles Freud's own psycho-physical parallelism:

> It is probable that the chain of physiological events in the nervous system does not stand in a causal connection with the psychical events. The physiological events do not cease as soon as the psychical ones begin; on the contrary, the physiological chain continues. What happens is simply that, *after a certain point of time,* each (or some) of its links has a psychical phenomenon corresponding to it. Accordingly, the psychical is a process parallel to the physiological—a dependent concomitant. (Freud, 1891, p. 55, emphasis added)

Psychological and neurobiological phenomena run parallel insofar as they are two incomplete perspectives of the same underlying process. This position counters the critique that neuropsychoanalysis is bioreductive. Drawing *correlations* with neurobiology (as in dual-aspect monism or psycho-physical parallelism) is not the same as suggesting that neurobiology can replace psychoanalysis. Nor does neurobiological correlation suggest that the brain is more real—and therefore *causes*—the mind. Psychoanalysis and neuroscience are two epistemological lenses for studying the same ontological thing in nature.

This is why, in general, neuroscientists emphasize the necessity of behavioral or psychological variables for interpreting brain imaging data. fMRI alone cannot validly be used to infer a psychological process (Krakauer et al., 2017). A single brain area is involved in multiple psychological processes, and a single psychological process involves multiple brain areas. This is the principle of "dynamic localization" (Kaplan-Solms & Solms, 2002), in the tradition of the Russian neuropsychologist Alexander Luria (1947). Psychological processes do not occur *inside* neural structures; they emerge *between* constellations of neural activity. Thus, they are *emergent* phenomena running *parallel* to the *activity* of neural structures.

For example, a patient suffered a right hemisphere stroke and developed a constellation of symptoms: left-sided hemineglect (i.e., ignoring objects in the left side of space), left-sided paralysis (due to left-sided motor control being localized to the right motor cortex), anosognosia for hemiplegia (i.e., denial of his paralysis), spatial and constructional challenges, and a narcissistic stance toward the hospital staff. A patient with a similar lesion had similar visuospatial and motor symptoms, except she was extraordinarily depressed. Another patient (again with a similar lesion) was similar, save for a striking hatred of the paralyzed limb (misoplegia) and a demand that it be removed. These examples (simplified from Kaplan-Solms & Solms, 2002) demonstrate how a single brain region (in this case, the right perisylvian convexity) is (1) involved in multiple psychological processes and (2) how similar lesions can result in different clinical presentations.

Likewise, consider a psychological concept like libido. Libido is strikingly similar to what is described in affective neuroscience as the SEEKING system (Solms, 2012; see Chap. 6). The SEEKING system is a "goad without a goal," an objectless motivational force that propels the organism to engage with novelty in the world. SEEKING involves multiple neural structures, with source nuclei in the midbrain ventral tegmental area sending dopaminergic projections to the striatum in the basal ganglia which then propagate to motor and prefrontal cortex, with characteristic frequency oscillations (Alcaro & Panksepp, 2011). Here, a single psychological construct (libido) involves a constellation of neural areas, neurotransmitters, and functional patterns. Neuropsychoanalytic dynamic localization admits a tethering between neurobiology and psychoanalysis without a simple, one-to-one mapping.

Claiming that neuropsychoanalytic dialogue is bio-reductive therefore missteps on the epistemological bridge between neuroscience and psychoanalysis. Reading Solmsian dual-aspect monism to the letter avoids a hierarchical division between neuroscience and psychoanalysis (cf. psycho-physical parallelism) and instead promotes a dynamic localization between epistemologies. This minimizes the risk of reductionism (in either direction) because one cannot privilege one register over the other. Neuroscience and psychoanalysis are two perspectives of the same thing. One cannot replace psychoanalytic notions with neural references, as both disciplines offer complementary (objective and subjective) perspectives on the mental apparatus. Neuropsychoanalysis is a bridge-discipline that links the two phenomenological perspectives (dual-aspect) of neuroscience and psychoanalysis to paint a more thorough picture of the ontological thing in nature (monism).

A Threefold Movement: Building a Meta-Neuropsychology

If neuroscience describes a distinct epistemological perspective of this ontological mental apparatus, then how does it contribute to psychoanalysis? I suggest that the problems concerning epistemology (see Chap. 2) are due, in part, to a lack of distinction between phenomena (observed from particular epistemologies) and meta-psychology (or theory). For example, from the subjective perspective, one observes depression. Psychoanalytically, one can construct a meta-psychology that theorizes a withdrawal of libido (Freud, 1917). Withdrawal of libido is an *abstraction*—a meta-psychological notion that makes sense of the subjective phenomenon. Neuroscience does the same thing. From the objective perspective, one observes decreased dopaminergic activity along mesocortical-mesolimbic structures (Panksepp, 1998). Neuroscientifically, one conceptualizes a SEEKING system that aims at optimistic engagement with uncertainty, the lack of which corresponds to a depressive presentation.

Dual-aspect monism builds a unifying meta-neuropsychological framework in which abstractions from both psychoanalysis and neuroscience can dialogue. Through such dialogue, the laws governing both psychological and neurobiological phenomena can be discerned (Solms, 2020a). Such laws would exist on the same plane as Freud's pleasure principle, death drive, ego, superego, id, and so on. They have a metapsychological status insofar as they go beyond phenomenological appearances. Whereas neuroscience and psychoanalysis are two epistemologies (objective and subjective) that deal with certain types of phenomena, they contribute concepts to a common meta-neuropsychology that incorporates these phenomena. Knowledge generated in psychoanalysis and neuroscience is not restricted to each discipline. Such knowledge—insofar as it contributes to meta-neuropsychology—concerns the workings of the same mental apparatus that refracts in objective and subjective phenomena. But how does one relate concepts derived from these two fields?

I propose a "threefold" movement of neuropsychoanalytic dialogue based on dual-aspect monism as a "method" for developing a Lacanian neuropsychoanalysis (Dall'Aglio, 2021, 2022). It can be described as follows:

(1) Map psychoanalytic concepts onto neuroscientific space.
(2) Observe the relationships among psychoanalytic concepts in neural space.
(3) Return to psychoanalytic space to consider what theoretical and clinical possibilities arise from the links suggested in neural space.

For example, consider the Lacanian concept of a signifier that binds some degree of drive-excitation (or *jouissance*). When mapped onto neural space, the signifier can be understood as a motoric-phonemic element that is tagged by a dopamine spike when it surprisingly alleviates some tension. Dopaminergic marking grants the motoric trace "incentive sensitization"—that is, motivation and reinforcement for the trace to be activated again, separate from need-activation (Bazan & Detandt, 2013; see Chap. 7). While language is especially salient due to phonological ambiguity, this logic applies equally to motor actions and cognitive schemas in general (Dall'Aglio, 2023). Thereby, in neural space, one finds a convergence between language, actions, and cognitive schemas. One can return to psychoanalytic space to consider the possibility that clinical principles applied to signifiers in speech might also be applicable to body actions and cognitive rules (see Chap. 14).

For example, a patient told a dream in which someone was throwing "keys" at them. This patient was dodging them for a while but described how one came really close to her chest, holding her palm in front of her chest to illustrate the proximity. Later in the session, she spoke about "avoiding" her parents' questions about her career choice (which was elected over motherhood). She also spoke about her resolution to spend more time with friends, grab drinks, and so on. Several events occurred recently that made her feel that the opportunity was imminent despite her tendency to self-sabotage her own relaxation. When describing this, she made the same gesture with her palm in front of their chest and said: "the opportunities are weighing on me." I pointed out that she made the same gesture when describing the dream and asked how these issues might relate to the dream. This led to productive work with the dream and the patient's surprise, as she had thought the dream was not related to anything in her life.

One could just as well have focused on the signifiers and associations to the dream. I am not attempting to demonstrate a superior technique derived from neuropsychoanalytic thinking. All I wish to illustrate is how a threefold movement between psychoanalysis and neuroscience can open theoretical and clinical possibilities. Here, the link was not made via the linguistic signifier but via a body gesture treated like a signifier.

This approach avoids the epistemological critique that neuroscience cannot inform psychoanalysis. Clinical or theoretical thinking remains in the psychoanalytic sphere. Ideas or connections proposed in neuroscientific space are allowed to propose possibilities and considerations. It keeps clinical possibilities open, as the brain is not turned to as an interpretive codebook. Such a conceptual dialogue is possible because these concepts—here, the signifier, emotional salience, motor trace—exist in the same meta-neuropsychological framework.

Given the dialogic structure of neuropsychoanalysis, this threefold movement can be reversed. Put formulaically:

(1) Map psychoanalytic space with neural concepts.
(2) Observe the relationship among neural concepts in psychoanalytic space.
(3) Return to neural space to consider what possibilities arise from the links suggested in psychoanalytic space.

Consider the neuroscientific concept of homeostasis. When mapped onto psychoanalytic space, one can find a parallel with Freud's pleasure principle—alongside the problems with the pleasure principle described under the heading of the death drive. Returning to neural space, once can consider how the idea of a death drive—the fracturing of the pleasure principle—can enrich or nuance how homeostasis is considered in the brain. I detail these implications in Chap. 8 and below.

Principles of a Lacanian Neuropsychoanalysis

When this threefold movement is employed with Lacanian psychoanalysis—specifically Lacan's meta-psychology of real, imaginary, and symbolic—one can discern meta-neuropsychological principles that are foundational for dialogue between Lacanian psychoanalysis and neuroscience. These revolve around the concept of the real (developed in Chap. 4). For now, it suffices to say that the Lacanian real stands for the antagonism, rift, or dehiscence within any system of representation, the inability of that system to fully resolve itself without contradiction or anomalous surplus. A Lacanian neuropsychoanalysis makes the real central to its meta-neuropsychological system.

Transcendental Materialist Dual-Aspect Monism

To situate a Lacanian perspective on dual-aspect monism, I rely heavily on Adrian Johnston's philosophy of Transcendental Materialism. Transcendental materialism seeks to provide a materialist philosophy that can account for the emergence of more-than-natural phenomena (e.g., psychoanalytic subjects) from natural phenomena (e.g., neurobiology) alone. For Johnston, any materialist philosophical system that does not account for naturalism is not properly materialist. Foreclosing naturalism opens the slippery slope for sacralization that installs various unquestionable guarantees. God in religion is the standard example. Famously, Lacan (1964) prohibited phylogenetic questions on the emergence of language, restricting the concern of psychoanalysis to the subject's immersion in speech and culture—privileging the biblical "In the beginning was the Word." Johnston (2019) critiques Lacan's refusal here as an instance of sacralization stemming from a rejection of phylogenetic, evolutionary, or biological considerations.

Importantly, psychoanalysis is not the only field that sometimes slips into sacralization. Johnston critiques common scientific understanding (and understandings of science often held by psychoanalysts) as retaining a notion of capital-N Nature—that is, a Nature of fully determinate and causal laws. In this view of Nature, there is a mappable chain of events that can be identified and described with mathematical-physical laws that would fully account for all phenomena from miniscule quantum physics to macro-scale cosmology in a deterministic fashion. Such a Nature is seamless, without interruption or impasse. Indeed, this is the Nature that frightens critics of neuropsychoanalysis—the concern with bio-reductionism is a fear that psychological phenomena will be reduced to a capital-B Brain that fully explains and resolves psychological questions (see Chap. 2).

In opposition to capital-N Nature, Johnston's transcendental materialism prioritizes a "weak nature alone." This "weak nature" is not one of full causal determinacy but one of cracks, misfirings, and failures. Nancy Cartwright's (1999) philosophy of science paints a "dappled world" of science in which the contradictions between laws (e.g., quantum physics,

Newtonian classical physics, Einstein's general relativity) are not inadequacies of scientific explanation. Those laws are "really real" at those levels—what appears in the form of epistemological contradiction between laws is in fact touching upon something in nature itself (Žižek, 2009, 2020). Nature is not a sweeping totality but a hodge-podge montage of pieces that do not neatly fit together.

Capital-E Evolution is another specter of Nature critiqued by Johnston. Everyday language speaks of something that "evolved to do" this or that. However, evolution is not oriented toward some pre-determined teleological goal or endpoint. The evolutionary imperative of "survive and reproduce" is a low bar. Simply considering the number of inheritable diseases is enough to demonstrate the inadequacy of the law of survival and reproduction to usher in ideal evolutionary progress. It is a "weak law" of a "weak nature." Evolution works with *random* mutations in ever-changing environments. Evolutionary leaps occur at moments of catastrophe and existential threats to species (cf. punctuated equilibrium; Gould & Eldredge, 1972). Evolution does not work in a deterministic fashion; rather, it is shaped by short-circuits and catastrophes—misfirings of nature.

At the human level, Johnston invokes cerebral organization as an example of a weak nature view of the brain. Building on work by Antonio Damasio (2010), Terrence Deacon (2011), David Linden (2008), and Joseph LeDoux (1998), Johnston emphasizes that the brain is not a single unified system called the capital-B Brain. Rather, the brain is a system of systems, a collection of cell assemblies with their own processes. These systems evolved disproportionately, under different circumstances in evolutionary history, now sedimented like a hodge-podge, kludge-like, Frankenstein-esque Creature. An evolutionarily ancient reptilian brainstem is wrapped by a mammalian limbic system upon which sits the neocortex whose prefrontal lobes are greatly expanded in humans. Each system has its own agenda (e.g., brainstem concerned with autonomic homeostasis; neocortex concerned with the external environment) and distinct means of dealing with stimuli (e.g., declarative neocortical capacities; non-declarative subcortical processes). Nevertheless, they are forced to work together for the organism's survival.

3 A Philosophical Basis for a Lacanian Neuropsychoanalysis

Johnston reads neuroplasticity—the brain's capacity to change its connections based on experience—as further evidence for the "weak nature" of the brain. For Johnston, neuroplasticity does not merely indicate the adaptive potential for neuronal function. It indicates the radically incomplete nature of the brain. Neuroplastic effects—such as epigenetic modification of gene expression—are not secondary "add-ons" to a fundamental or "true" nature. Rather, the brain is *genetically determined to not be genetically determined* (Ansermet & Magistretti, 2007). The innate nature of the brain includes incompleteness, holes, and openings for the impact of experience. It is therefore impossible to think of the brain without the historical and present experiences that shape(d) it. A transcendental materialist view of the brain privileges the conflictual, short-circuiting, and incomplete nature of neural systems. Paradoxically, uniquely human features of subjectivity arise precisely from such short circuits and antagonisms.

Before detailing a transcendental materialist perspective on Solmsian dual-aspect monism, I wish to clarify that I am not seeking to replace dual-aspect monism. Nor am I attempting to account for the emergence of the division between objective and subjective epistemological registers. Rather, I hope to extend Solmsian dual-aspect monism to include the notion of antagonism-in-nature so essential to transcendental materialism.

Recall that critics highlight the epistemological gap between neuroscience and psychoanalysis as cause for dismissing any productive potential from neuropsychoanalytic dialogue (see Chap. 2). Beyond proposing that a common meta-neuropsychology can offer a theoretical space for *considering* implications of such dialogue, I also propose that this epistemological chasm is not the only gap to be reckoned with. In line with transcendental materialism, I propose that the ontological level—the mental apparatus—is itself antagonistic. There is antagonism-in-nature, the self-sundering of natural substance. This necessitates theorizing an ontology of antagonism and dehiscence which would refract *within* both neuroscientific and psychoanalytic epistemologies. This nuances Solmsian dual-aspect monism by suggesting meta-neuropsychological bridges not only at points of conceptual similarity but also at points of contradiction or failure. The meta-neuropsychology built by a transcendental materialist dual-aspect monism would furthermore include the conflict between these perspectives as part of its own system.

Slavoj Žižek provides a precise formula for a transcendental materialist dual-aspect monism:

> the only proper reply to this challenge [of the brain sciences in the face of psychoanalysis] is to meet the brain sciences' neuronal Real [antagonism within the brain] with another Real [antagonism in the mind], not simply to ground the Freudian *semblant* within the neuronal Real. In other words, if psychoanalysis is to survive and retain its key status, *we have to find a place for it within the brain sciences themselves, starting from their inherent silences and impossibilities.* (Žižek, 2009, p. 177, emphasis in the original)

Where traditional dual-aspect monism draws correlations between neuroscientific and psychoanalytic concepts (e.g., SEEKING and libido; Solms, 2012; see Chap. 6), transcendental materialist dual-aspect monism would also privilege connections between points of contradiction or short-circuiting. These would be instances of the Lacanian real—specifically, the opening of one real (e.g., neuronal) onto another (e.g., psychoanalytic).

Specifically, I propose that such "meetings of reals" are the nodal points of emergence of more-than-natural phenomena from natural substance alone (Johnston, 2019). These would be the "certain point[s]" after which psycho-physical parallelism emerges for Freud (1891). An ontology of antagonism allows the emergence of more-than-natural phenomena not from the deterministic laws of the previous level, but from the *failures* of those laws to fully regulate or assert themselves. Such failures create more-than-natural phenomena that are irreducible to the prior level. Or, they are only reducible to the prior level insofar as that level self-sunders. Chapter 8 develops this point regarding neuronal antagonism and the emergence of affective consciousness.

Therefore, where Solmsian dual-aspect monism seeks to identify the laws that govern the mental apparatus and explain both biological and psychological phenomena, transcendental materialist dual-aspect monism permits a lawful reductionism that includes the failure of these laws. Asserting the weakness of these laws to perfectly implement themselves allows a materialist philosophy to account for the emergence of unique dimensions of human subjectivity from a weak nature alone.

A transcendental materialist dual-aspect monism thus sets a philosophical basis for a specifically Lacanian neuropsychoanalysis. Here, I propose principles for a Lacanian neuropsychoanalysis that are rooted in the ontology of antagonism in transcendental materialism. In addition, I will discuss how they overcome the criticisms that have been levied against (Lacanian) neuropsychoanalysis (see Chap. 2).

There is no Homeostasis of Homeostasis[1]

One of Lacan's famous aphorisms is *There is no Other of the Other*. One interpretation of this statement is that it amounts to saying the Other is barred, that there is no external guarantee for the consistency of the symbolic. There is no second Other who can guarantee the operation of the (first) Other as not lacking (Johnston, 2005).

Following Jonathan Lear (2000), Johnston (2018) argues that Freud's (1920) assertion of a principle beyond the pleasure principle acts similarly to positing an Other beyond the Other. In *Beyond the Pleasure Principle*, Freud grapples with various clinical problems that demonstrate the inadequacy of the pleasure principle to explain psychic phenomena. To solve this problem, Freud posits an additional principle—the death drive—to make sense of the failures of the pleasure principle. In a way, this guarantees (the strength of) the pleasure principle. The pleasure principle works; there is just another drive toward death that comes in the way.

Instead, for Johnston, death drive names the inadequacy of the pleasure principle itself. As Lear puts it: "what lies 'beyond the pleasure principle' isn't another principle, but a lack of principle" (Lear, 2000, p. 85). The pleasure principle fails to fully assert itself to adequately manage tension in the mental apparatus, with the problematic of earlier experiences of satisfaction that yearn for repetition. There is no beyond the pleasure principle; the pleasure principle is barred.

---

[1] I gratefully adopt this phrase from Ariane Bazan (personal communication).

When the pleasure principle is mapped onto homeostasis, one thereby arrives at the formula: *There is no Homeostasis of Homeostasis* (Ariane Bazan, personal communication). As I will develop in Chap. 8, the brain may very well obey a homeostatic imperative to minimize tension. However, multiple homeostatic systems exist in a conflictual relationship with each other. There is no inherited genetic algorithm for regulating conflicts among homeosta*ses*. No meta-homeostasis exists that can resolve problems of homeostases.

The principle *There is no Homeostasis of Homeostasis* therefore avoids criticisms that Lacanian neuropsychoanalysis returns psychoanalysis to a purely homeostatic model. Homeostasis *is* a key principle in neurobiology. However, rather than reject homeostasis (and biology with it), Lacanian neuropsychoanalysis highlights how a rigorous neuroscientific study of homeostatic systems reveals the insufficiency of homeostasis— not as a limitation of scientific knowledge but as a feature of the mental apparatus itself. The concept self-sunders, betraying a point of silence in the deterministic capacity of neuroscientific laws (Žižek, 2009).

There is no Cerebral Relationship: Or, There is Cerebral Non-Rapport

A further principle takes up Lacan's aphorism *There is no sexual relationship*. Lacan consistently opposes supposed totalities, unities, and complementarities. In this case, Lacan rejects the idea of a perfect harmony between the sexes. Lacan's diction (the French *rapport* is typically translated as *relationship* but can also be translated as *ratio* in the mathematical sense) specifies that there is no (symbolic) *formula* for connecting two sexed subjects that does not leave some (real) remainder unaccounted (Copjec, 2015).

As noted above, Johnston (2019) reviews several neuroscientists who highlight the sedimented, out-of-joint nature of evolutionarily dehiscent brain systems. Playing on Lacan's statement, Johnston asserts *There is no intra-cerebral relationship*. That is, there is no *formula* for a perfect rapport among brain systems that eliminates all inconsistency, lack, or surplus. Using Damasio's metaphor, different brain systems form an "odd-couple,"

3 A Philosophical Basis for a Lacanian Neuropsychoanalysis

a strange marriage among systems that must cooperate for the survival of the organism. This could be phrased otherwise: *The Instinct does not exist*. Capital-I Instinct—in the sense of a perfect reflex arc that eliminates tension—is a simplified biological strawman. Chapter 7 details features of the brain's functional architecture that render Instincts structurally prone to warping and aberrancy. For now, it suffices to say that Lacanian neuropsychoanalysis does not replace psychoanalytic drive with seamless instinct.

However, one can go a step further. Copjec (2015) highlights that the difference *between* the sexes is not the only incongruity discerned by Lacan (1972–1973) in his formulas of sexuation. A contradiction exists *within* each side of the formulas—that is, there is a contradictory logic within the masculine formulas as well as the feminine formulas. These details are beyond the scope of this chapter. It suffices to say that Lacan's *There is no sexual relationship* does not designate an incompatibility between two simply existing entities. Each side is itself rift with an antagonism that bears consequences for desire and enjoyment.

In a Hegelian light, Žižek (2020) interprets Lacan's aphorism not only as a claim of non-existence (of the sexual relationship) but also an assertion of the existence of a negativity: *There is sexual non-rapport*. I use the term "negativity" as the opposite of "positivity" to designate something whose presence exerts itself but cannot be grasped in a positivist logic; it can only be grasped through failure, contradiction, or surplus. It seeks to capture an existential judgment without predicative judgment—simply put, the assertion of existence without being able to describe what it is (Copjec, 2015). This is a way of conceptualizing the Lacanian real (see Chap. 4).

With Žižek's *There is sexual non-rapport*, not only is there no formula for the sexual relationship (i.e., no relation of drive-satisfaction that completely eliminates tension—see Chap. 4). There *exists* a (Hegelian) determinate negation—the dimension of negativity that is not only destructive but also formative. To be clear, this is not the claim of a positive (substantial) existence; it is an affirmation of a negative. With more precise Lacanian terminology, one would say *There ex-sists sexual non-rapport* (where "exists" designates positivist existence and "ex-sists" indexes the insistence of a negativity). In other words, the real is given a primacy and motor force rather than a simple absence. The ex-sistence of the sexual non-rapport is a *present absence* with structural and clinical consequences for the human subject.

Transposing this reasoning onto the brain, I claim *There is cerebral non-rapport*. That is, the contradictions and conflicts among brain systems are not only evidence for the lack of a higher-order, unifying logic (e.g., a meta-Homeostasis). There is a *causal* power of the real—or the Žižekian meeting of reals—that has determinate effects on subjectivity (Last, 2021). Specifically, cerebral non-rapport has the effect of producing more-than-natural phenomena out of the brain alone (Deacon, 2011; Johnston, 2019; Linden, 2008).

These two principles—*There is no Homeostasis of Homeostasis* and *There is cerebral non-rapport*—demonstrate how Lacanian neuropsychoanalysis not only attunes to the real but also makes the real central to its theoretical system. A corollary here is that Lacanian neuropsychoanalysis does not rely on normative statistical logic as a *goal*. Rather, it *recognizes* normativity (what is the Lacanian Name-of-the-Father if not the *normative* delusion for knotting the registers?) while centralizing the *supplementary* logic of insistent negativity that disrupts any equalization or homeostatic formula. It is insufficient to argue that the brain knows nothing of the real because neuroscience uses fMRI. The image of the brain is not the final "truth" for neuroscience. This is an unfortunate (and fantasmatic) oversimplification of neuroscience. As I will demonstrate in Chap. 8, Lacanian neuropsychoanalysis can demarcate points of contradiction that carve out a real immanent to symbolic operation at the level of innate affective systems. This real has the causal dimension of affective consciousness as felt uncertainty.

Provoking Prediction Error

This final principle addresses the ethical criticism of clinical normativity. Lacanian neuropsychoanalysis privileges the cracks, inconsistencies, and surpluses produced out of immanent antagonisms within the functioning of the brain. One such concept—from computational neuroscience—is the notion of *prediction error*. I will develop these details in Chaps. 5, 6, and 7. For now, it suffices to say that prediction errors signal that things are not going as planned for how the brain anticipates its engagement with the world. Affective arousal is a key class of prediction errors indicating deviations from homeostatic set-points. Prediction errors are *surprising*. They drive changes in the brain, enabling adaptation to situations of uncertainty.

If we begin from the notion that homeostasis is impossible—*There is no Homeostasis of Homeostasis*—then clinical neuropsychoanalysis cannot hope to eliminate prediction error. Clinical technique recognizes that surprise not only drives neural change (memory reconsolidation; Beckers & Kindt, 2017; Schroyens et al., 2017); surprise also marks the real. Encounters with the real are indicated by surprise, regardless of affective valence (Soler, 2015). Technique thereby orients toward surprise as the space of the divided subject.

Lacanian interventions aim to provoke surprise, to highlight the unexpected, and draw attention to the new (Fink, 2011; Israely, 2018; Lacan, 1964). Slips are clinically significant because they are not what the subject intended or expected to say. Word-play draws salience to phonological ambiguity and the unexpected possibilities therein. Scansion—the unexpected ending of a session—cuts the patient's speech to disrupt standard narratives or emphasize something without allowing the patient to cover their tracks with more speech.

Clinical applications of Lacanian neuropsychoanalysis are detailed in Chap. 14. At this point, the orientation toward negativity—and the correlative surprise—as a clinical method indicates how Lacanian neuropsychoanalysis in no way smuggles a normative or educational ethics into the clinic. Such criticisms conflate neuroscientific discourse with clinical practice and fail to carefully read how neuroscientific ideas might be used to inspire ideas in the clinic. Moreover, insofar as clinical practice is first and foremost the domain of psychoanalysis, dialogue with neuroscience would not simply graft scientific ethics onto psychoanalysis. Neuropsychoanalysis is a space to consider possibilities, including clinical technique and formulation, afresh through interdisciplinary dialogue.

These principles of a Lacanian neuropsychoanalysis—*There is no Homeostasis of Homeostasis*, *There is cerebral non-rapport*, and *Provoking prediction error*—guide the meta-neuropsychological integration of

Lacanian psychoanalysis with computational neuroscience, affective neuroscience, and neuropsychology detailed in the rest of this book. They are put forward *contra* criticisms of bio-reductionism and normalization often levied against neuropsychoanalysis (see Chap. 2). Throughout this book, I will show how these principles open a perspective on the brain that fits with the neuroscientific data, does not do epistemological violence to integration with Lacanian psychoanalysis, and—most importantly—opens new avenues for theoretical and clinical work.

Indeed, Darian Leader advocates for such interdisciplinary work to counter the tendency to reify psychoanalytic ideas. Contemporary psychoanalysis should seek to "continue the project that…[Lacan] had opened"—specifically, the critical formalization and re-thinking of the concepts in Freud's oeuvre as linked to psychoanalytic practice (Leader, 2021, p. 134). As Fred Guterl put it regarding neuropsychoanalysis, "it's not a matter of proving Freud wrong or right, but of finishing the job" (Guterl, 2002). It is in this spirit that I present a Lacanian neuropsychoanalysis.

References

Alcaro, A., & Panksepp, J. (2011). The SEEKING mind: Primal neuro-affective substrates for appetitive incentive states and their pathological dynamics in addictions and depression. *Neuroscience & Biobehavioral Reviews, 35*(9), 1805–1820. https://doi.org/10.1016/j.neubiorev.2011.03.002

Ansermet, F., & Magistretti, P. (2007). *Biology of freedom: Neural plasticity, experience, and the unconscious.* (S. Fairfield, Trans.). Other Press.

Bazan, A., & Detandt, S. (2013). On the physiology of jouissance: Interpreting the mesolimbic dopaminergic reward functions from a psychoanalytic perspective. *Frontiers in Human Neuroscience, 7*, 709.

Beckers, T., & Kindt, M. (2017). Memory reconsolidation interference as an emerging treatment for emotional disorders: Strengths, limitations, challenges, and opportunities. *Annual Review of Clinical Psychology, 13*, 99–121.

Cartwright, N. (1999). *The dappled world: A study of the boundaries of science.* Cambridge University Press.

Copjec, J. (2015). *Read my desire: Lacan against the historicists* (2nd ed.). Verso.

Dall'Aglio, J. (2021). What can psychoanalysis learn from neuroscience? A theoretical basis for the emergence of a neuropsychoanalytic model. *Contemporary Psychoanalysis, 57*(1), 125–145. https://doi.org/10.1080/00107530.2021. 1894542

Dall'Aglio, J. (2022). Neuropsychoanalysis: What, how, and why. In G. Gargiulo & J. Turtz (Eds.), *Enriching psychoanalysis: Integrating concepts from contemporary science and philosophy* (pp. 119–146). Routledge. https://doi.org/10.432 4/9781003271499-11

Dall'Aglio, J. (2023). Extending the theory of premature automatization: The fantasy as an abstract rule in hierarchical cognitive control. *Neuropsychoanalysis, 25*(1), 27–42. https://doi.org/10.1080/15294145.2023.2183888

Damasio, A. (2010). *Self comes to mind: Constructing the conscious brain*. Pantheon/Random House.

Deacon, T. (2011). *Incomplete nature: How mind emerged from matter*. Norton.

Fink, B. (2011). *Fundamentals of psychoanalytic technique: A Lacanian approach for practitioners*. Norton.

Freud, S. (1891). *On aphasia: A critical study* (E. Stengel, Trans.). International Universities Press.

Freud, S. (1917). Mourning and melancholia. In *The Standard Edition of the Comlpete Psychological Works of Sigmund Freud, Vol. XIV* (J. Strachey, Ed., Trans.) (pp. 237–258). Hogarth Press.

Freud, S. (1920/1955). Beyond the pleasure principle. In *The standard edition of the complete psychological works of Sigmund Freud, Vol. XVIII* (J. Strachey, Ed., Trans.) (pp. 1–64). Hogarth Press.

Gould, S., & Eldredge, N. (1972). Punctuated equilibria: An alternative to phyletic gradualism. In T. Schopf (Ed.), *Models in Paleobiology* (pp. 82–115). Freeman Cooper.

Guterl, F. (2002, November). What Freud got right. *Newsweek, 140*(20), 50–51. https://www.newsweek.com/what-freud-got-right-142575

Israely, Y. (2018). *Lacanian treatment: Psychoanalysis for clinicians*. Routledge.

Johnston, A. (2005). *Time driven: Metapsychology and the splitting of the drive*. Northwestern University Press.

Johnston, A. (2018). *A new German idealism: Hegel, Žižek, and dialectical materialism*. Columbia University Press.

Johnston, A. (2019). *Prolegomena to any future materialism, Volume two: A weak nature alone*. Northwestern University Press.

Kaplan-Solms, K., & Solms, M. (2002). *Clinical studies in neuro-psychoanalysis: Introduction to a depth neuropsychology* (2nd ed.). Karnac Books.

Krakauer, J., Ghazanfar, A., Gomez-Marin, A., MacIver, M., & Poeppel, D. (2017). Neuroscience needs behavior: Correcting a reductionist bias. *Neuron, 93*(3), 480–490. https://doi.org/10.1016/j.neuron.2016.12.041

Lacan, J. (1964/1978). *The seminar of Jacques Lacan, Book XI: The four fundamental concepts of psychoanalysis* (J.-A. Miller, Ed., A. Sheridan, Trans.). Norton.

Lacan, J. (1972–1973/2000). *The seminar of Jacques Lacan, Book XX: On feminine sexuality, the limits of love and knowledge* (J.-A. Miller, Ed., B. Fink, Trans.). Norton.

Last, C. (2021). The difference between neuroscience and psychoanalysis: Irreducibility of absence to brain states. *Neuropsychoanalysis, 23*(1), 27–38. https://doi.org/10.1080/15294145.2021.1926312

Leader, D. (2021). *Jouissance: Sexuality, suffering and satisfaction*. Polity.

Lear, J. (2000). *Happiness, death, and the remainder of life*. Harvard University Press.

LeDoux, J. (1998). *The emotional brain: The mysterious underpinnings of emotional life*. Simon & Schuster.

Linden, D. (2008). *The accidental mind: How brain evolution has given us love, memory, dreams, and God*. Harvard University Press.

Luria, A. (1947). *Traumatic aphasia: Its syndromes, psychology and treatment*. The Hague.

Panksepp, J. (1998). *Affective neuroscience: The foundations of human and animal emotions*. Oxford University Press.

Schroyens, N., Beckers, R., & Kindt, M. (2017). In search for boundary conditions of reconsolidation: A failure of fear memory interference. *Frontiers in Behavioral Neuroscience, 11*, 65. https://doi.org/10.3389/fnbeh.2017.00065

Soler, C. (2015). *Lacanian affects: The function of affect in Lacan's work* (B. Fink, Trans.). Routledge.

Solms, M. (2012). Are Freud's "erogenous zones" sources or objects of libidinal drive? *Neuropsychoanalysis, 14*(1), 53–56. https://doi.org/10.1080/15294145.2012.10773688

Solms, M. (2020a). New project for a scientific psychology: General scheme. *Neuropsychoanalysis, 22*(1–2), 5–35. https://doi.org/10.1080/15294145.2020.1833361

Solms, M. (2020b). Response to the commentaries on the "New Project". *Neuropsychoanalysis, 22*(1–2), 97–107. https://doi.org/10.1080/15294145.2020.1843215

Solms, M., & Turnbull, O. (2002). *The brain and the inner world: An introduction to the neuroscience of subjective experience*. Other Press.

Žižek, S. (2009). *The parallax view*. MIT Press.

Žižek, S. (2020). *Sex and the failed absolute*. Bloomsbury.

Part II

The Enjoying Brain

4

The Concept of *Jouissance*

Abstract This chapter develops the Lacanian concept of *jouissance*—a traumatic excess of enjoyment which may not be felt as such—in relation to the real, imaginary, and symbolic registers. I demonstrate how Lacan extracts these concepts (and others, such as *das Ding*, *objet a*, $, S1, S2) from some of Freud's major texts. I also discuss how the Lacanian real is an antagonism *immanent* to the symbolic, the symbolic's own structural inconsistency. This conceptualization will be central to my subsequent integrations with neuroscience.

Keywords Real • Imaginary • Symbolic • Jouissance • Lacan • Freud • Instinct • Drive • Affect • Mirror stage

Here I describe the meta-psychology of *jouissance* as indexed to the real, imaginary, and symbolic registers. There are many paradigms of *jouissance* in Lacanian theory (Leader, 2021; Miller, 2020). I will specifically develop

---

Here I summarize arguments from Dall'Aglio (Dall'Aglio, 2021).

© The Author(s), under exclusive license to Springer Nature Switzerland AG 2024   **43**
J. Dall'Aglio, *A Lacanian Neuropsychoanalysis*, The Palgrave Lacan Series,
https://doi.org/10.1007/978-3-031-68831-7_4

the idea of *jouissance* as excess because, in my view, it offers the greatest conceptual enrichment to neuroscience and neuropsychoanalysis. That said, my elaboration of the concept is one perspective among many.

Freud and the Problem with the Pleasure Principle

Throughout Freud's body of work, he relies heavily on the pleasure principle to explain the operations of the mental apparatus (Freud, 1895, 1900, 1905, 1911, 1920). Under the pleasure principle, increases of tension are felt as unpleasure, and decreases of tension are felt as pleasure. For Freud, some stimulus—most notably, unmet drive-needs—generates tension. The pleasure principle drives the subject to discharge this tension to the lowest possible state. It is a logic of tension-reduction. Although the reality principle tolerates delays to immediate discharge and inhibits immediate release, it ultimately aims to minimize tension. The reality principle thereby supplements the pleasure principle's aim.

On the other hand, Freud repeatedly encounters clinical phenomena that lead him to question this model. One of the first challenges is the problem of infantile sexuality, described in *Three Essays on the Theory of Sexuality* (Freud, 1905; Schuster, 2016). Infantile sexuality—with its partial (component) drives—seeks a strange enjoyment in the *increase* of tension. Stimulation of the erotogenic zones (e.g., oral and anal orifices) and fore-pleasure are exemplary. Freud discerns an enjoyment derived from excess—distinct from the pleasure associated with tension-reduction.

Despite this puzzling enjoyment, Freud downplays the problem of infantile partial drives by privileging the organizing role of the Oedipus complex and genital sexuality. One can compare the original 1905 publication of the *Three Essays* with subsequent editions of the text. Freud gradually scaffolds a psychosexual developmental model (i.e., oral stage, anal stage, phallic stage, latency, genital stage) around infantile sexuality. This development toward the genital stage shies away from the disturbing primacy of infantile partial drives. The partial drives are united under the flag of genital sexuality which is oriented toward the decrease of tension

(Schuster, 2016). Freud props up genital sexuality as the teleological endpoint of libidinal development, effectively privileging the pleasure principle over the infantile partial drives.

Nevertheless, Freud continues to recognize uncanny fractures within the pleasure principle. He groups many of these phenomena—the negative therapeutic reaction, repetition, masochism, nightmares, and so on—under the heading of the death drive (Freud, 1920, 1924). Lacan takes this contradiction seriously, questioning the primacy of genital sexuality in contrast to the excess-seeking logic of infantile sexuality. How can the partial drives, aiming toward the *increase* of tension, be united under the harmonious goal of *decreasing* tension? Lacan thereby rejects genital sexuality as the goal or model for the drive (Lacan, 1959–1960, 1964).

Jouissance is indexed to excess. Importantly, *jouissance* names a specific enjoyment *in the tension itself*. Such enjoyment is distinct from increasing tension for the sake of subsequent relief. The tension-seeking quality of infantile sexuality and partial drives is characteristic of *jouissance*. Some element of infantile sexuality—forever "polymorphously perverse"—always persists and disturbs the subject's tension-minimizing efforts.

This sentiment can be detected in Freud's musing that

> Sometimes one seems to perceive that it is not only the pressure of civilization but something in the nature of the [sexual] function itself which denies us full satisfaction and urges us along other paths. (Freud, 1930, p. 105)

Obstacles to full satisfaction (the pleasure principle's regime of total reduction of drive-tension) do not principally derive from cultural prohibitions or social rules epitomized by the Oedipus complex. *Drive itself* poses an obstacle to the pleasure principle—specifically, the tendency to seek an excess enjoyment (*jouissance*) that "urges us along other paths" (Dolar, 2017; Johnston, 2005).

Furthermore, *jouissance* links repetition to sexual enjoyment. Žižek gives a striking example:

> imagine someone shakes my hand as a gesture of greeting, but then he goes on, continuing to rhythmically squeeze my hand for no obvious reason—I would certainly experience this continuous squeezing as an indication that

something "dirty" is going on, that an unwelcome sexualization is taking place. (Žižek, 2020b, p. 209)

Repetition invokes sexualization, as Freud notes in "The 'Uncanny'" (Freud, 1919). This is especially so when there is repression: repetition "for no obvious reason." Linking *jouissance* to the enjoyment of sexualized repetition places *jouissance* in the logic of the death drive. For Lacan,[1] death drive dictates endless repetition that aims at the impossible return to an original traumatic irruption of tension. At the core of the psyche exists a disturbing, traumatic drive-tension that cannot be represented or bound, thereby stultifying the regime of the pleasure principle (Lacan, 1964; Soler, 2015).

Tracing the Contours of the Real: *Das Ding* and *Objet a*

Here, Lacan's register theory (real, symbolic, imaginary) is clarifying. The symbolic can be thought of as the system of representations: Freudian *Vorstellungen*, ideas, mnemic traces. The imaginary can be thought of as attempts at or formations of wholeness, unity, or understanding. For Lacan (1954–1955), the ego is the exemplar of the imaginary—insofar as the ego delineates contours of the body, gives a sense of self, proffers understanding, and so on. Symbolic and imaginary both involve representing experience, binding it in some way, and giving it rooting or form.

On the other hand, the real is what drops out of representation, what *resists* representation. As Lacan put it: "The real is what resists symbolization absolutely" (Lacan, 1953–1954, p. 66). It is *non-representational* but nevertheless *insists*. I use the term *negativity* (as a negation of positivity) to characterize the real because, for Lacan, the real has no positive (i.e., substantial) status (Žižek, 2020b; Zupančič, 2017). The dream's navel is exemplary. Some element in experience insists insofar as its absence

---

[1] One should acknowledge that the death drive is understood differently in Lacanian literature (see Zupančič, 2017). Most notably, the Lacanian reading privileges the "repetition compulsion" and self-effacing or negating elements of the death drive over Freud's musings on a drive toward biological death.

imposes itself as resisting capture in representation or meaning (Freud, 1900). The original traumatic irruption of the drive that breaks the "stimulus-barrier" is another example (Freud, 1920). In this sense, the Lacanian real has a traumatic status.

Tracing how Lacan extracts his notion of the real from Freud will help elucidate this admittedly difficult concept, especially as it sits relative to the symbolic, imaginary, and *jouissance*. Lacan (1959–1960) returns to Freud's (1895) *Project for a Scientific Psychology* to find there a connection between *jouissance* and negativity. When describing the development of speech, Freud writes:

> Speech innervation is originally a path of discharge…it is a portion of the path to *internal change*, which represents the only discharge till the *specific action* has been found. This path acquires a secondary function from the fact that it draws the attention of the helpful person (usually the wished-for object itself) to the child's longing and distressful state; and thereafter it serves for *communication* and is thus drawn into the specific action. At the start of the function of judgement, when the perceptions, on account of their possible connection with the wished-for object, are arousing interest…their complexes are dissected into an unassimilable component (the Thing [*das Ding*]) and one known to the ego from its own experience (attribute, activity)—what we call understanding. (Freud, 1895, p. 366, emphases in the original)

Several pieces deserve note here. First, the experience of internal drive-tension leads the child to cry out—creating an internal change *prior to communicating* with an object (i.e., the mother or primary caregiver). In other words, the drive does not have an innate, pre-given object. Second, the motor action—crying out, which later becomes speech—has an initial function of *internal change*; communication is a secondary function of speech. Here, one can detect Lacan's (1955–1956) distinction between *signifier* as the motoric-phonemic element of speech and the *signified* as the conveyed meaning.

Moreover, when the mother does appear, there is a split between what can be understood and represented ("known to the ego" as "attribute") and *das Ding*, an "unassimilable component." The "function of judgment"—whether regarding attribution of good and bad or regarding existence (Freud, 1925)—is thereby made with reference to *re-finding* an object. The subject constructs its representational world in relation to a logically prior object that was not grasped and must be re-found. Lacan proposes:

> Everything...that is articulated as good or bad divides the subject...irrepressibly, irremediably, and no doubt with relation to the same Thing [*Ding*]. There is not a good and a bad object; there is good and bad, and then there is the Thing. The good and bad already belong to the order of the *Vorstellung* [signifier], they exist as clues to that which orients the position of the subject [in relation to *das Ding*]. (Lacan, 1959–1960, p. 63)

Lacan's split here differs from Melanie Klein's (1946) splitting of good and bad part-objects. These good and bad part-objects belong to the order of attributive judgment and, therefore, the domain of the symbolic (i.e., *Vorstellung*). Such things can be known, grasped, and imaged by the ego. Lacan demarcates a more original split in Freud's *Project*: between representation and *das Ding*. *Das Ding* is not a neglected perception. It is a negativity that emerges simultaneously with representation.

Freud (1905) describes a similar situation in relation to the drive:

> At a time at which the first beginnings of sexual satisfaction are still linked with the taking of nourishment, the sexual instinct has a sexual object outside the infant's own body in the shape of his mother's breast. It is only later that the instinct loses that object, just at the time, perhaps, when the child is able to form a total idea of the person to whom the organ that is giving him satisfaction belongs. As a rule the sexual instinct [drive] then becomes auto-erotic. (Freud, 1905, p. 222)

Once more, representation—the "total idea of the person"—coincides with a loss. A split emerges between representation and a dimension of loss. Importantly, the lost object was never grasped (i.e., represented, imaged) to begin with. In a certain sense, the object exists only as a lost object, a negativity.

These discussions allow us to draw a distinction between *das Ding* and *objet a* (described in detail below). They also draw out two dimensions of *jouissance* concerning trauma and repetition. Here I rely on theoretical developments by Copjec (2004), Žižek (2020a, 2020b), and Zupančič (2017). I refer the interested reader to these authors for more detailed discussion.

Das Ding can be thought of as the original, traumatic irruption of negativity. It indexes the overwhelming unknown of the Mother, the ultimate Thing to which the infant is subjected in its prematurity and motor helplessness. It is a dangerous excitation insofar as the infant cannot yet discharge it through speech. *Das Ding* corresponds to traumatic, overwhelming *jouissance*, the irruption of excessive drive tension that perturbs the homeostasis of the mental apparatus.

When the caregiver intervenes, the cry becomes tethered not only to communication but also to representation. For instance, the mother might interpret the child's cry as "You're hungry." Such interpretation binds some of the drive's tension and allows some degree of discharge (Verhaeghe, 2004). Put simply, the mother holds the baby in a certain fashion, speaks in soothing motherese, and provides milk. Various actions, images, and words are tied to the drive-tension, and the provision of nourishment allows some alleviation (Van de Vijver et al., 2017). This constellation constitutes the experience of satisfaction (Freud, 1895).

To the extent that motor actions and words (i.e., signifiers) assuage drive-arousal, the unnamable component becomes transposed into the representational field (Verhaeghe, 2004). In other words, the traumatic non-representational *jouissance* of *das Ding* becomes titrated by words and images. Although these representations manage some of the tension and allow discharge, for Lacan, some lack or negativity persists. The originally sought for object remains impossible because it did not exist in a positive, substantial sense in the first place. But now, the subject has actions, words, and images (representations) to go about trying to re-find this object. *Objet a* names this lack in the field of representation.

Some examples will be helpful here. Michael Myers, the antagonist of the movie *Halloween* (Green, 2018), wonderfully exemplifies *das Ding*. Darian Leader remarks on 2018 remake:

> The journalists are eager to understand what happened all those years ago, to give a meaning and motivation to the murders, yet we know that Myers has no said a single word throughout his imprisonment. In his silence, he incarnates an opacity of meaning, and the question of 'What did he want?' To produce some sort of reaction, which they hope will generate speech, one of the visitors confronts Myers with the Halloween mask he had worn when he committed his murders. As the mask is revealed, the courtyard is filled with a commotion, the other inmates screaming and crying, yet Myers remains absolutely silent. (Leader, 2021, p. 66)

Myers attains a status of terror through silence, an unspoken density surrounded by screams and a desire for speech (i.e., representation), However, the search for clarity runs up against a horrifying unknown. The violent homicides further support the dread tethered to Myers's silence.

Contrast this traumatic concretization of the real with *The Ambassadors* (1533), the famous painting by Hans Holbein the Younger of two adorned figures with a distorted, slanted object in the foreground. When looking directly at the painting, the object remains unclear. Only when standing at a far edge of the room at a certain angle can the viewer see that the object is a skull (this painting technique is called anamorphosis). Importantly, when the skull becomes visible, the rest of the painting is distorted.

This example perfectly illustrates *objet a*. *Objet a* names the lack or torsion *within* representational space—something enigmatic and unknown that disturbs the seamlessness of the image. Here, an enigmatic object is included within the painting. When shifting perspective to see this object, the rest of the image is distorted. *The Ambassadors* does not form a totality without some ambiguous tension whose positive location cannot be pinpointed (Lacan, 1964).

Lacan offers another example of *objet a*: the story of Zeuxis and Parrhasios. Parrhasios painted a veil on a wall, "a veil so lifelike that Zeuxis, turning towards him said, *Well, and now show us what you have painted behind it*" (Lacan, 1964, p. 103, emphasis in the original). Of course, nothing is painted behind the veil, but this "nothing" (i.e., negativity) is precisely what arouses the interest of Zeuxis. *Objet a* exists as an enigmatic ambiguity or negativity whose excess can only be alluded to by

4 The Concept of *Jouissance*   51

representation, "a certain bulge in the phenomenal veil" (Lacan, quoted in Soler, 2015, p. 24).

Consider the following clinical example. A patient presented with a history of physical abuse by a teacher. She spoke in tears when recounting the abuse, to which she attributes much of her current struggles (various post-traumatic stress symptoms, agoraphobia, and so on). She had fought for the school and police to investigate the issue, but no one believed her. The lack of acknowledgment of her story added gravity to the trauma.

A few sessions later, she spoke about attending a parent-teacher conference for her children. She had to sit near several teachers and felt disturbed. Although she described the teachers as friendly and welcoming, she could not shake a residual anxiety: "One of them could be dangerous. Maybe not, but I don't know." There was something about the teachers—more so in this meeting than other moments where she had to be around teachers—that struck a nerve, although she could not name it. To express her frustration during the meeting, she was somewhat oppositional, glaring, disagreeing, and so on.

These instances approximate the difference between *das Ding* and *objet a*. *Das Ding* takes on a traumatic, imposing force—not only the weight of childhood physical abuse but also the vacuum in which her speech received no answer or recognition. It is a disturbing, traumatic kernel. The later encounter with teachers did not have the same weight, but there remained some potentiality (an operative yet unknown element, *objet a*) that roused her suspicion: "could be…Maybe…I don't know."

These examples also demonstrate two dimensions of *jouissance*. The traumatic encounter of *das Ding* indexes an overwhelming excitation before which the child is helpless—motor activity and speech are of no avail. *Objet a* orients toward the excesses and uncertainties within representation: the unknown possibility captures her attention; anxiety, anger, and fearful avoidance disturb her peace.

One can further tether this distinction to two types of repetition described by Lacan (Lacan, 1964; Zupančič, 2017). The first is repetition as typically understood, the repetition of this or that action or scene. This patient repeatedly complained about frustrating authority figures (parents, coaches, bosses, and so on) who frustrated her for no apparent reason. The hostility with which she described these figures betrayed some strange enjoyment in directing her hatred toward these objects. Here is a titrated *jouissance* in the repeated situation.

However, Lacan makes a further, unintuitive claim regarding repetition: "Repetition demands the new" (Lacan, 1964, p. 61). One can understand this as the death drive's aim to return to the original traumatic irruption that broke the stimulus-barrier (Freud, 1920). In seeking to return to the representation-shattering event, the drive aims at precisely what is unknown and not represented—hence, "new." Here, drive procures a more intense (perhaps traumatic) enjoyment in the encounter with novelty that drifts toward *das Ding*.

The Lacanian Drive

Here, it will be fruitful to describe the Lacanian model of the drive. Lacan takes seriously Freud's contention that the drive is a concept

> on the frontier between the mental and the somatic, as the psychical representative of the stimuli originating from within the organism and reaching the mind, as a measure of the demand made upon the mind for work in consequence of its connection with the body. (Freud, 1915a, p. 122)

Drive is not a unity—Lacan calls it a *"montage"* (Lacan, 1964, p. 169). Freud describes it as a frontier-concept: drive straddles the frontier between body and mind, containing the division within itself. Drive-source and drive-pressure originate in the non-representational body, the "demand made upon mind for work." On the other hand, drive is tethered to representation; it is a psychical *representative* involving actions and objects.

Recall how Freud (1905) describes the emergence of the drive in relation to representation ("the total idea of the person") and object loss. The full representation of the object of satisfaction does not exist; some loss always accompanies the experience of satisfaction. Therefore, the drive's aim for satisfaction is an impossible demand, insofar as no object grasped can match the primal experience of satisfaction (Van de Vijver et al., 2017).

Drive-source and drive-pressure sit in an uncomfortable kludge with the drive-aims (actions marked to achieve this or that mode of satisfaction) and drive-objects. Johnston (2005) places drive-source and drive-pressure on the side of the real, incessantly insisting on the repetition of an original enjoyment that cannot be represented. No actions or objects—on the side of symbolic-imaginary representations—eliminate the tension. Hence, the drive exerts a constant pressure (Freud, 1915a). Insofar as drive consists of these four components (source, pressure, aim, and object) the montage of the drive itself contains the antagonism of what cannot be grasped or eliminated by the symbolic or imaginary (Johnston, 2005). Drive itself is divided—the drive's lack is not physiological but immanent to the structure of drive itself. And nevertheless, this tension provides the motor force for psychical life, the demand for work—which, for Lacan, is a demand for speech (Copjec, 2004; Fink, 2011).

One can link these points regarding drive to the distinction between *das Ding* and *objet a* and between traumatic and titrated *jouissance*. *Das Ding* can correspond with the traumatic irruption of tension (cf. the death drive's breaking of the stimulus barrier; Freud, 1920), the excess of drive-source and drive-pressure. Insofar as speech metabolizes some of this tension, certain actions—the "specific action" (Freud, 1895)—are marked as pathways along which the drive might seek to (re)find satisfaction (Van de Vijver et al., 2017). Certain representations—motor aims and objects—are re-sought because they approximate the initial experience of satisfaction and allow some management of drive tension (*jouissance*). In the process of marking certain actions and representations (i.e., signifiers), *jouissance* "sticks" to these signifiers (Zupančič, 2017). These signifiers repeat for two reasons: (1) they mark the contours of the original drive-irruption and (2) there is some enjoyment (*jouissance*) in the repetition itself that fails to re-find the object. For Lacan (1964), the object of the drive is a lack forever ungrasped: *objet a*.

Insofar as some degree of sense and wholeness is achieved, *jouissance* is likewise invested in experiences of totality or understanding. The quintessential Lacanian example is the infant, wrought with motoric incapacities and disparate component drives, who sees the total image of itself in the mirror. This is Lacan's mirror stage (Lacan, 1953–1954, 1954–1955). A jubilant assumption of the image as the basis of the ego corresponds with a libidinal investment in the image—that is, *jouissance* is caught up in the image. One can expand this sense of the imaginary beyond images alone to any instances of totality, wholeness, or understanding and the corresponding enjoyment associated with it (Fink, 2011; Leader, 2021; Soler, 2015).

Nevertheless, such repeated signifiers and egoic crystallizations are doomed to fail, because there was no full representational grasping of this initial encounter (Freud, 1895, 1905; Lacan, 1959–1960). Some lack persists despite the tension worked over in the subject's motoric and representational capacities. The drive's path thus encircles what is "not quite it" in the myriad objects, actions, words, and images it encounters. Drive revolves around *objet a*, procuring enjoyment not in the encounter but in the repeated (failed) attempt to return to the original missed encounter (Lacan, 1964). Certain key signifiers mark the contours around *objet a*.

Here one should distinguish drive and desire, especially since Lacan (1964) calls *objet a* the "object-cause of desire." One desires—is motivated to work, to produce, to create, to *speak*—because of the drive's structure that contains some irresolvable antagonism (i.e., the real). The persistent negativity or lack (*objet a*) causes the ongoing movement of desire—the seeking of this or that object which, once obtained, does not exactly scratch the itch, and thus we wish for something else. Quite simply, we go on desiring because we never quite get what we want. To fully satisfy desire would extinguish desire. In this sense, desire is never satisfied, being continually displaced onto what one does not possess. Think of the child who makes endless demands of the mother: attention, food, toys, water, and so on. Desire emerges precisely in the rift between (impossible) demand and the particular objects that fail to meet it (Lacan, 1964). On the other hand, drive *always* derives enjoyment, insofar as the very repetition of the failure to meet impossible demand is itself enjoyed. There is a *jouissance* in failed repetition, the failure to grasp the unknown.

Recall the repeated hatred and fear the patient held toward authority figures in the aforementioned clinical example. This constellation of negative affects motivates a series of desires—her own wish to be a proper mother, her wish to put them in their place, fantasies of saving others,

4 The Concept of *Jouissance* 55

and so on. One can detect a strange enjoyment in the repetition of these negatively valenced emotions. This is *jouissance* in the repetitive movement of the drive's circuit. The desire to remove the traumatic tension remains unsatisfied, which is why an endless series of particular objects can take the manifest position.

A further example helps illustrate this. Consider the film *Kung Fu Panda* (Stevenson & Osborn, 2008). A clumsy panda named Po dreams of becoming a Kung Fu master. A prophecy pronounces him the famed Dragon Warrior who will defeat an evil, brutal leopard named Tai Lung who seeks revenge. Po must prove himself worthy so his master will give him the Dragon Scroll, an ancient text containing the greatest Kung Fu knowledge. Tai Lung, who also seeks the Dragon Scroll, battles with Po over its possession.

Žižek (2011) describes how the Dragon Scroll operates in the position of *objet a* in the film. The Dragon Scroll stands for complete mastery—a symbolic system in which no lack exists or a perfect state of complete drive satisfaction. Such a promise of power draws the desire of both Po and Tai Lung. Nevertheless, when Po finally obtains the scroll and opens it, he finds nothing but a mirror reflecting his image.

To grasp this crucial moment, one must reject the commonsense wisdom that Po himself is the secret knowledge or that he possessed it all along. This would be an instance of fetishist disavowal: I recognize there is no ultimate answer, but I nevertheless believe there is one. The *objet* of the drive and the promise of total tension-reduction is a lure, a void that lacks positive, substantial existence. It is only materialized in the symbolic, representational world as an enigma or absence. No knowledge of total Kung Fu mastery (or drive resolution) exists, yet this very lack motivates the drive's demand for work and the subject's desire. Enjoyment resides precisely in the humorous fighting (i.e., tension, search for aims and objects) to obtain the Scroll.

The Real: Antagonism Immanent to the Symbolic

The Unconscious is Structured like a Language

With these conceptual bearings, one can better grasp Lacan's aphorism that "the unconscious is structured like a language." At the most basic level, the unconscious is structured as a network of associations: mnemic traces or representatives—that is, signifiers. The interplay of these signifiers produces effects: meanings, significations, affects, and so on. Consider the interpretive logic Freud describes early on for dreams, symptoms, jokes, and repression: the primacy of switch-words, dreams as rebuses, phonological similarity and ambiguity, and so on (Bazan, 2011). At this level, interpreting the childhood origins of a phobia, for instance, through its associations can resolve the symptom. Here is the symbolic dimension of the unconscious (Lacan, 1964; Verhaeghe, 2002).

However, the Lacanian unconscious can be pushed further—demonstrating equally what is particular to Lacan's theory of language. Lacan highlights the distinction in Freud's meta-psychology between primal and secondary repression (Freud, 1915b, 1915c). Secondary repression—"repression proper"—is the rendering unconscious of this or that representation that held a (pre)conscious cathexis. It corresponds to the operations of the symbolic unconscious: displacement of affect onto associated ideas, transformation into anxiety, conversion into bodily innervation, expansion into a phobic network or obsessional labyrinth, and so on. However, primal repression refers to an original repression of a trace which *never* had a (pre)conscious cathexis to begin with. The trace subject to primal repression—the drive-representative—was not known and then rendered unconscious. It is an original repression that exerts an attractive pull upon associated ideas and continually opposes becoming conscious through an anti-cathexis.

Here, in the structure of the unconscious, is an *internal externality*. At the core of the unconscious is an original repression that opposes (via an anti-cathexis) the surrounding representatives subject to repression proper. A limit point, linked to the drive (the primally repressed drive-representative), resides at the core of the unconscious that structurally resists representation (cannot be rendered conscious) yet structures the network of representations around it. Recall again the navel of the dream that resists interpretation (Freud, 1900). This is the unconscious as real (Lacan, 1964; Verhaeghe, 2002).

Language, for Lacan, has the same structure. There are signifiers, representations that can be spoken, forgotten, remembered, dreamed, bungled, and so on. Yet, every saying contains some unsayable (Copjec, 2004). Some alluded to excess insists as resisting full expression. Meaning remains ever unstable and potentially evolving, and no sentence quite says it all. Hence, the unconscious is structured like a language. Both the unconscious and language have the structure of some real that perturbs any harmonious equalization or resolution in the (symbolic) system.

Jouissance, S1, S2

We can now situate some additional Lacanian terms with greater precision. The real names the traumatic unknown gap within the symbolic system of representations. *Jouissance* is the excess excitation of the drive (cf. *das Ding*) that irrupts at the point of the real.

Before the pressure of the real, the subject turns to actions and speech (representations, the symbolic) to manage some of the tension. Certain key actions and representations that manage some drive-tension are marked in this process. *Jouissance* sticks to these signifiers, marking and charging them. These signifiers are *master signifiers*: "S1" in Lacanian algebra. Such signifiers resonate with the weight of *jouissance* and nod toward the real of the drive (Zupančič, 2017). They demarcate the contours of the real, now transposed into the uncertainties, excesses, and lacks in the domain of action and speech (cf. *objet a*).

Lacan calls an S1 a signifier all alone, not linked up to other signifiers (Lacan, 1964, 1972–1973). Meaning emerges in the process of signifiers being connected to each other (Lacan, 1954–1955). An S1 does not operate at the level of meaning or understanding, yet it nevertheless organizes the symbolic network insofar as the representative indexing the real (Zupančič, 2017).

For example, a patient spoke about hating the phrase "put yourself first" in his childhood. His mother consistently encouraged him to prioritize his own goals and successes, often over friendships and relationships. He said "I never knew what that meant. Aren't my friendships and relationships also for me?" Surely putting yourself first had some line or limit—as was implicitly communicated in the form of family responsibilities, expectations, and so on. What "first" exactly meant puzzled him. He spoke about conflicts between achievement-striving, relationships, how much to push himself, guilt, and so on.

In this example, the signifier "first" can be thought of as an S1. He remained puzzled and did not know precisely what it meant or—more specifically—what it would look like for him in his life. And yet, it has some operative value insofar as the question that arose around the S1—what does "first" mean?—is precisely what he lives out. He himself is surprised by his energy and motivation, pointing to the drive (*jouissance*) marked by the guiding idea of "put yourself first."

Here one can also observe the linking up of a signifier all alone (S1) with other signifiers (abbreviated as "S2" in Lacanian algebra). Linking signifiers generates knowledge, meaning, and signification. Consider how meaning emerges in the relationship between words that can be differentiated from each other—adding words allows one to grasp and manipulate meaning (Lacan, 1955–1956).

In the example of "first" as S1, "first" is a master signifier that indicates how this person organized *jouissance*. But what does it mean to put yourself first? How does this patient's "first" manifest in his life? Linking S1 to S2 (the battery of signifiers) is revealed in the myriad particular manifestations of putting himself first (and questioning how to do this), balancing accomplishment, career-driven work, relationships, and so on. In other words, situations are staged such that some sense can be attributed to "first"—despite its chronic opacity.

A signifier is thereby not restricted to the symbolic. S1 indicates the dimension of the signifier tethered to the real. It is bound to *jouissance*, eludes meaning, and (can be) cut off from associated representations. S2 points to the dimension of the signifier on the side of the symbolic: the differential, associative linking of representations in chains. Meaning and understanding can be produced through this signifier-organization, which would be the level of the imaginary.

Excess and Negativity: The Divided Subject ($) and *Objet a*

As is evident from the above discussions, the Lacanian real is spoken about as both excess and negativity. How can one reconcile this? Žižek (2009) suggests that these are two points on the same moebius strip. From the perspective of the *structure* of the symbolic, there is a core negativity or antagonism, a disruption to the system of representations (cf. an original repression or loss). From the perspective of the *elements* in the system, however, this negativity is registered as an excessive (yet unknown) piece: *objet a*.

This also allows us to situate two additional mathemes: the divided subject (barred-subject, $) and *objet a*. For Lacan, no signifier represents the subject fully. The subject emerges as divided *between* signifiers; the subject as desiring emerges in the interplay of signifiers which *insufficiently* attain satisfaction (Israely, 2018; Verhaeghe, 2004). The subject is thus divided and lacking: $. In the representational world, *objet a* is the objectal correlate of $, the elusive surplus that escapes capture in representation yet orients the drive (Žižek, 2009). If $ is the divided subject of negativity, *objet a* is the surprising excess of *jouissance*. In more banal terms, we desire because we are not fully appeased (i.e., we are lacking, barred, $), and whatever we obtain is not quite "it" (i.e., not *objet a*). Thus we go on desiring (as lacking, as $): *objet a* is the cause of desire.

Desubstantializing the Real

Importantly, for Lacan, the real is *desubstantialized*. Here, Lacan departs further from Freud. The unconscious (as real) is not a thing or a place buried deep within the psyche; it is a *formal* gap or contradiction (Lacan, 1964, 1972–1973). It is not a hard kernel that resists representation. From this perspective, *das Ding* as the Thing, or an original drive-representative primally repressed, is theoretically misleading. These ideas suggest some hard substance—a Thing—that traumatically intrudes and forever eludes the grasp of the symbolic and imaginary.

For Lacan, there is no outside of the symbolic. The real is not some-Thing external to the symbolic. Rather, the real *names the structural antagonism of the symbolic itself*. Language is incomplete, structurally lacking the signifier that would resolve all meanings and nail down understanding (Zupančič, 2017). The symbolic contains antagonisms immanent to its operative logic; representation is structurally destabilized (Verhaeghe, 2004; Žižek, 2020b). While there are different ways to position oneself with respect to structural antagonism (Copjec, 2015), the real persists as a contradiction immanent to the symbolic order.

The notion of real as some-Thing external to the symbolic suggests a certain (quasi-Freudian) deterministic view of causality. Some-Thing happens, some trauma for instance, that perturbs the symbolic (an irruption of *jouissance* due to a traumatic encounter) and has effects such as symptoms. In other words, the real as substantial kernel intrudes and warps symbolic space.

Shifting to the real as structural antagonism of the symbolic itself changes this view of causality. The symbolic itself is destabilized as incomplete. There exist points of antagonism where *jouissance* emerges—*jouissance* is the excess excitation arising at points of *structural* instability. This is the level of *impossibility*. Beginning from impossibility (of complete resolution, for instance), one's specific life experiences and encounters are *contingent* features for concretizing and dealing with *jouissance*. Such contingencies, however, become subject to the repetition compulsion and become *necessary*. There is no substantial-real that warps symbolic space; the symbolic is itself structurally warped and antagonistic. Here, causality is not deterministic. It is a causality of the real-as-antagonism *immanent* to structure—a causality that begins with impossibility that opens contingency which becomes necessity (Lacan, 1964; Lacan, 1972–1973).

From the perspective of necessity (e.g., the endless repetition of a certain symptom), one might look back and retroactively image [sic] the specter of a horrible *Ding* whose traumatic *jouissance* caused the symptom. However, from the perspective of real as structural antagonism (impossibility), this is a secondary move that obscures the deadlock of the symbolic. For example, Johnston (2005) highlights that Oedipal prohibition of total *jouissance* (in Freudian terms, prohibiting the mother) retroactively creates the fantasy that total *jouissance* was once possible but has been given up. The particular details of the Oedipal constellation are contingent yet become the determinant (necessary) features of the neurosis. However, Oedipal prohibition veils the impossibility of such *jouissance* in the first place. Even if the child could fully have the mother, full satisfaction would not occur because even the mother is not the Mother, the real Thing (Žižek, 2020a).

This allows us to make a further point regarding Lacanian criticisms that neuroscience cannot capture the real (see Chaps. 2 and 3). With this view of the real, the real is not some Thing to be captured or not; it is not a substance. The real is a negativity *immanent to symbolic logic*. One can thereby seek to "capture" the real in neuroscience by identifying points of structural antagonism within neuroscientific theories of the brain. It is to these theories that I now turn.

References

Bazan, A. (2011). Phantoms in the voice: A neuropsychoanalytic hypothesis on the structure of the unconscious. *Neuropsychoanalysis, 13*(2), 161–176. https://doi.org/10.1080/15294145.2011.10773672
Copjec, J. (2004). *Imagine there's no woman: Ethics and sublimation* (2nd ed.). MIT Press.
Copjec, J. (2015). *Read my desire: Lacan against the historicists* (2nd ed.). Verso.
Dall'Aglio, J. (2021). Sex and prediction error, part 1: The metapsychology of jouissance. *Journal of the American Psychoanalytic Association, 69*(4), 693–714. https://doi.org/10.1177/00030651211042000
Dolar, M. (2017). Of drives and culture. *vInternational, 1*(1), 55–79.
Fink, B. (2011). *Fundamentals of psychoanalytic technique: A Lacanian approach for practitioners*. Norton.
Freud, S. (1895/1966). Project for a scientific psychology. In *The standard edition of the complete psychological works of Sigmund Freud, Vol. 1* (J. Strachey, Ed., Trans.) (pp. 281–391). Hogarth Press.
Freud, S. (1900/2010). *The interpretation of dreams* (J. Strachey, Ed. & Trans.). Basic Books.

Freud, S. (1905/1955). Three essays on the theory of sexuality. In *The standard edition of the complete psychological works of Sigmund Freud, Vol. II* (J. Strachey, Ed., Trans.) (pp. 123–246). Hogarth Press.

Freud, S. (1911/1958). Formulations on the two principles of mental functioning. In *The standard edition of the complete psychological works of Sigmund Freud, Vol. XII* (J. Strachey, Ed., Transs) (pp. 213–226). Hogarth Press.

Freud, S. (1915a/1957). Instincts and their vicissitudes. In *The standard edition of the complete psychological works of Sigmund Freud, Volume 14* (J. Strachey, Ed., Trans.) (pp. 109–140). Hogarth Press.

Freud, S. (1915b/1957). Repression. In *The Standard edition of the complete psychological works of Sigmund Freud, Vol. XIV* (J. Strachey, Ed., Trans.) (pp. 141–158). Hogarth Press.

Freud, S. (1915c/1957). The unconscious. In *The standard edition of the complete psychological works of Sigmund Freud, Vol. XIV* (J. Strachey, Ed., Trans.) (pp. 159–215). Hogarth Press.

Freud, S. (1919/1955). The uncanny. In *The standary edition of the complete psychological works of Sigmund Freud, Vol. XVII* (J. Strachey, Ed., Trans.) (pp. 217–256). Hogarth Press.

Freud, S. (1920/1955). Beyond the pleasure principle. In *The standard edition of the complete psychological works of Sigmund Freud, Vol. XVIII* (J. Strachey, Ed., Trans.) (pp. 1–64). Hogarth Press.

Freud, S. (1924/1961). The economic problem of masochism. In *The standard edition of the complete psychological works of Sigmund Freud, Vol. XIX* (J. Strachey, Ed., Trans.) (pp. 155–170). Hogarth Press.

Freud, S. (1925/1961). Negation. In *The standard edition of the complete psychological works of Sigmund Freud, Vol. XIX* (J. Strachey, Ed., Trans.) (pp. 233–240). Hogarth Press.

Freud, S. (1930/1961). Civilization and its discontents. In *The standard edition of the complete psychological works of Sigmund Freud, Vol XXI* (J. Strachey, Ed., Trans.) (pp. 57–146). Hogarth Press.

Green, D. (Director). (2018). *Halloween* [Motion Picture].

Holbein the Younger, H. (1533). *The ambassadors* (oil on oak). London: National Gallery.

Israely, Y. (2018). *Lacanian treatment: Psychoanalysis for clinicians*. Routledge.

Johnston, A. (2005). *Time driven: Metapsychology and the splitting of the drive*. Northwestern University Press.

Klein, M. (1946). Notes on some schizoid mechanisms. *The International Journal of Psychoanalysis, 27*, 99–110.

Lacan, J. (1953–1954/1991). *The seminar of Jacques Lacan, Book I: Freud's papers on technique* (J.-A. Miller, Ed.; J. Forrester, Trans.). Norton.
Lacan, J. (1954–1955/1991). *The seminar of Jacques Lacan, Book II: The ego in Freud's theory and in the technique of psychoanalysis* (J.-A. Miller, Ed.; S. Tomaselli, Trans.). Norton.
Lacan, J. (1955–1956/1997). *The seminar of Jacques Lacan, Book III: The psychoses* (J.-A. Miller, Ed., R. Grigg, Trans.). Norton.
Lacan, J. (1959–1960/1992). *The seminar of Jacques Lacan, Book VII: The ethics of psychoanalysis* (J.-A. Miller, Ed., & D. Porter, Trans.) Norton.
Lacan, J. (1964/1978). *The seminar of Jacques Lacan, Book XI: The four fundamental concepts of psychoanalysis* (J.-A. Miller, Ed., A. Sheridan, Trans.). Norton.
Lacan, J. (1972–1973/2000). *The seminar of Jacques Lacan, Book XX: On feminine sexuality, the limits of love and knowledge* (J.-A. Miller, Ed., B. Fink, Trans.). Norton.
Leader, D. (2021). *Jouissance: Sexuality, suffering and satisfaction*. Polity.
Miller, J.-A. (2020). Six paradigms of jouissance. (J. Haney, Trans.). *Psychoanalytical Notebooks, 34,* 11–77.
Schuster, A. (2016). *The trouble with pleasure: Deleuze and psychoanalysis*. MIT Press.
Soler, C. (2015). *Lacanian affects: The function of affect in Lacan's work* (B. Fink, Trans.). Routledge.
Stevenson, J., & Osborn, M. (Directors). (2008). *Kung Fu Panda* [Motion Picture].
Van de Vijver, G., Bazan, A., & Detandt, S. (2017). The mark, the Thing, and the object: On what commands repetition in Freud and Lacan. *Frontiers in Psychology, 8,* 2244. https://doi.org/10.3389/fpsyg.2017.02244
Verhaeghe, P. (2002). Lacan's answer to the classical mind/body deadlock: Retracing Freud's Beyond. In S. Barnard & B. Fink (Eds.), *Reading seminar XX: Lacan's major work on love, knowledge, and feminine sexuality* (pp. 109–120). SUNY Press.
Verhaeghe, P. (2004). *On being normal and other disorders: A manual for clinical psychodiagnostics* (S. Jottkandt, Trans.). Other Press.
Žižek, S. (2009). *The parallax view*. MIT Press.
Žižek, S. (2011). *Living in the end times* (Revised ed.). Verso.
Žižek, S. (2020a). *Hegel in a wired brain*. Bloomsbury.
Žižek, S. (2020b). *Sex and the failed absolute*. Bloomsbury.
Zupančič, A. (2017). *What IS sex?* MIT Press.

5

The Free Energy Principle

Abstract The Free Energy Principle, as developed by Karl Friston, has been a cornerstone of Mark Solms's recent work in neuropsychoanalysis. All self-organizing systems aim to minimize free energy, but how do they do this? In this chapter, I outline the basic conceptual arguments of the Free Energy Principle as applied to biological self-organizing systems and the brain. This includes concepts like free energy, Markov blankets, predictions, prediction error, and precision. The Free Energy Principle offers crucial conceptual resources in the shift to *informatic* uncertainty which open bridges with Lacanian psychoanalysis.

Keywords Self-organizing • Predictive coding • Prediction • Prediction error • Markov blanket • Neuroscience • Bayesian • Friston • Active inference • Neuropsychoanalysis

The Free Energy Principle (Friston, 2010) is a physical theory that describes the existence and operation of any "thing" that can be distinguished (in the statistical sense) from another "thing." It is a theory of self-organizing systems that distinguishes "system" from "not-system." As developed by Karl Friston, the Free Energy Principle (and its application in terms of *active inference* and the *Bayesian brain hypothesis* in neuroscience; see below) has been proposed as a unified theory of neural functioning, where all neural processes can be conceptualized as aiming to minimize free energy (elaborated below). While the Free Energy Principle offers striking explanatory power for a range of neural functions (Friston, 2010; Parr et al., 2022) and has garnered substantial empirical support (for review, see Hodson et al., 2024), it is not without criticism nor is it universally accepted by the neuroscientific community. However, its extensive use by Mark Solms makes an elaboration of the theory necessary before progressing to its neuropsychoanalytic implementation.

Before discussing the Free Energy Principle, I will present a brief case (Mr. A) which I will reference in subsequent chapters to illustrate certain neuropsychoanalytic and computational concepts.

Mr. A began therapy struggling with depression, anxiety, and stress. He worked as a retail manager. He felt unfulfilled with his job, having majored in economics in college. He spoke scornfully about lacking the "professional connections" to obtain positions in the economics industry. That being said, he devoted considerable attention to detail in his retail management job, dividing tasks into component parts and addressing them accordingly.

As therapy progressed, it became clearer that this strategy of "separating" (which was his approach to problems in general, not only work) had two origins: his education in economics (he was quite taken by utility-theory and optimization approaches) and an encounter with a man in his childhood. When he was four, he saw a snake at the zoo. It triggered an unbearable panic. A man turned to him and said: "if you're feeling afraid, just separate that thing or feeling from your mind so you can focus on other things."

He also complained of relationship problems. His husband was decried as misunderstanding him and not giving him enough attention. Mr. A did everything in his power to please his husband: cooking, giving him rides, paying for expensive trips—all in hopes of more attention. Mr. A racked his brain to come up with solutions to his relationship difficulties, trying to "optimize" each "variable" (e.g., his needs, his social desires, his husband's needs, his husband's family, and so on).

Two general principles slowly emerged in the treatment. Mr. A recognized that "I strive for the best" because of a fear of being "inferior." He also admitted: "I sabotage myself." He complained about not doing enough for himself, getting in his own way, and so on.

One session, he talked about an interaction he had with his husband. His husband just returned from a religious retreat to Japan. He said to Mr. A that he should take up contemplative practices (e.g., mindfulness, meditation, yoga, etc.) because he found it so helpful. Mr. A found himself enraged and felt that his husband was giving unwanted "input"—especially after leaving him alone for several weeks. However, Mr. A revealed none of this to his husband and instead quipped that he might try it. In the session, he reflected that it was "useless" to reply from his "silly" perspective (i.e., his feeling angry) because his husband would take it up as Mr. A being "frustrated" and "needy" and then respond in kind with anger.

Self-Organizing Systems

Karl Friston (2010) proposes the Free Energy Principle as a unifying theory of neural functioning. Here, I describe this computational neuroscientific model, synthesizing accounts by Friston (2010), Solms (2021), and Parr et al. (2022). I will not provide mathematical details, as these are beyond my expertise. I refer the interested reader to the aforementioned authors for deeper elaboration.

Self-organizing systems, such as biological systems, must resist the thermodynamic tendency to entropy. A self-organizing system is any "thing" that can be distinguished from something else ("not-system"). Resisting entropy is a corollary to avoiding dissipation of this distinction between system and not-system.

This dissipation can be characterized as a form of entropy. *Free energy* is a measure of entropy—or informatic "surprise" (more below)—in a system. The Free Energy Principle states that self-organizing systems minimize their free energy through Bayesian inference. All functions in the system can thereby be described as minimizing free energy through (approximate) Bayesian inference. Let us consider the building blocks for this separation and the process of Bayesian inference in free energy minimization.

For a system to distinguish itself from its environment, it must create a partition. This is called a "Markov Blanket." A Markov blanket includes all elements that divide the system from the not-system, the set of variables whose values must be known to determine the state of the not-system. It is a conditional barrier that cuts the system off from the not-system and allows the former to resist dissipation. A simple example is the cell membrane of an amoeba, where the amoeba is the self-organizing system.

A Markov blanket contains two states: active states and sensory states. These refer to whether the system is acting on the environment (i.e., active state; say the amoeba pushes out its pseudopods to obtain nutrients) or whether the system is being affected by the environment (i.e., sensory state; the cell membrane detects a protein binding to a receptor on the membrane). This is why the Markov Blanket is a conditional barrier: it separates the system from the not-system while still allowing reciprocal impacts from the system to the not-system and vice versa.[1]

The necessity of a Markov blanket for a self-organizing system to exist introduces two crucial points. First, recall that the self-organizing system's existence depends on resisting entropic dissipation. This means that the self-organizing system must maintain certain states despite a changing environment. A cell, for example, must maintain a certain level of sodium, temperature, and so on. If these bounds are violated, the system ceases to be a system (e.g., proteins denature, sodium concentrations deteriorate cellular processes, etc.). Complex mammalian organisms have further homeostatic parameters (see Chap. 6). These are "preferred" or "desired states"—states that the system aims to remain in.

The probability of states viable for the system must therefore have low entropy: the likelihood of the system staying in viable states must be high; the probability of the system occupying other (less or non-viable states) must be low. Deviations from preferred states are less likely and therefore more surprising—meaning there is greater free energy, greater entropy. (Mathematically, free energy is an upper bound on surprise; see

---

[1] Of note, this means that self-organizing systems are not "closed systems."

below.) In terms of the above example, there must be a high probability that the cell remains within certain bounds of sodium levels. Minimizing free energy thereby corresponds to minimizing surprise. As Friston puts it:

> Here, 'a fish out of water' would be in a surprising state (both emotionally and mathematically). A fish that frequently forsook water would have high entropy. Note that both surprise and entropy depend on the agent: what is surprising for one agent (for example, being out of water) may not be surprising for another. Biological agents must therefore minimize the long-term average of surprise to ensure that their sensory entropy remains low. (Friston, 2010, p. 127)

A second consequence of the Markov blanket is that the system is not in direct contact with the not-system. External states only impact the system *via* sensory states of the Markov blanket. This creates a problem: how does the system "know" whether these external states are surprising? Friston elaborates:

> So far, all we have said is that biological agents must avoid surprises to ensure that their states remain within physiological bounds…But how do they do this? A system cannot know whether its sensations are surprising and could not avoid them even if it did know. This is where free energy comes in: free energy is an upper bound on surprise, which means that if agents minimize free energy, they implicitly minimize surprise. Crucially, free energy can be evaluated because it is a function of two things to which the agent has access: its sensory states and a recognition density that is encoded by its internal states (for example, neuronal activity and connection strengths). The recognition density is a probabilistic representation of what caused a particular sensation. (Friston, 2010, p. 128)

In other words, the self-organizing system forms an internal model—called a "generative model"—of its environment. Equipped with an internal model and sensory states of its Markov blanket, the system can measure the degree of entropy by comparing its generative model (of the environment) with sensory input from the environment. This is formalized as (variational) free energy. Importantly, this model is informatic, rather than substantial:

This (variational) free-energy construct was introduced into statistical physics to convert difficult probability-density integration problems into easier optimization problems. It is an information theoretic quantity (like surprise), as opposed to a thermodynamic quantity…In the present context, free energy provides the answer to a fundamental question: how do self-organizing adaptive systems avoid surprising states? They can do this by minimizing their free energy. (Friston, 2010, pp. 128–129)

Let us return to the fact that the system is not in direct contact with the external world due to the Markov blanket. Here enters Bayesian inference. Because of the Markov blanket's partition, the system's generative model of the environment is formed through *inferences* of external states. These inferences are called "predictions." A prediction is a probability distribution representing the cause of sensory inputs. Predictions formalize the notion of "beliefs"—the system's generative model is its *beliefs* about hidden (unknown) external states. This has the further consequence that an agent aims to form a model of its environment. Minimizing free energy is thus equivalent to maximizing the evidence for the internal model, which corresponds to maximizing evidence for the existence of the agent. Hence, self-organizing systems are called "self-evidencing."

Predictions involve both sensory and active states of the Markov blanket. In the context of more complex self-organizing systems—especially mammals and the human brain—sensory and active states involve slightly distinct dynamics. For sensory states, predictions infer the causes of sensory input. Predictions (the generative model of the environment) are compared with sensory input. The difference between the two is called "prediction error." Sensory prediction errors drive model updating, adjusting predictions to minimize prediction errors. In the case of sensory states, this minimizes variational free energy.

More specifically, Bayesian probability separates predictions into "prior predictions" and "posterior predictions." Prior predictions are compared to sensory input, generating prediction errors. Prediction error drives updating of the predictive system, forming posterior predictions that better minimize prediction error. Perception is the process of updating the internal model to minimize (sensory) prediction error.

For active states, predictions select inputs to match the internal model (prior predictions of sensory states). Actions are selected on the basis of expected free energy. In other words, how much prediction error is expected from current action given current beliefs about the world? Actions with the lowest expected free energy are selected. Sensory states are then registered, and prediction errors drive either model updating (perception) or further actions to reduce error signals. This is the action-perception cycle. Action and perception are therefore two sides of the same coin of minimizing free energy.

Predictions and Prediction Errors in the Brain

According to the Free Energy Principle, the brain is constantly involved in a process of updating predictions to minimize prediction error. Neuronal activity and connectivity form a generative model, inferences about the (hidden) external world. Interoceptive (proprioception, blood pressure monitoring, hunger detection, etc.) and exteroceptive (vision, audition, somatosensation, etc.) sensory systems form the sensory states of the Markov blanket while thinking (i.e., mental action) and skeletal musculature form active blanket states.

In the brain, the generative model is organized hierarchically, spanning basic sensory inferences (e.g., luminance, color, line orientation, motion) to more complex, abstract inferences (e.g., others' intentions). This corresponds to the hierarchical organization of cortical and subcortical structures. For example, perceptual brain regions are hierarchically organized from unimodal perceptual regions, to heteromodal representational capacities, to amodal, abstract processes.[2] Predictions are passed down from higher regions to lower regions to "explain" incoming inputs; unexplained inputs are passed upward as error signals. This reciprocal

---

[2] Unimodal sensory cortex refers to cortex involved in processing information in one sensory modality (e.g., just vision). Heteromodal association cortex refers to regions involved in integrating information from multiple modalities (e.g., vision, audition, and somatosensation) to form a re-representation of perceptual objects. Amodal association cortex refers to regions involved in processing information abstracted beyond concrete perceptual qualities (e.g., semantic information, cognitive control, etc.). See Kaplan-Solms and Solms (2002) and Solms and Turnbull (2002) for an accessible and psychoanalytically-friendly introduction to brain organization.

connectivity finds biological implementation in the structure of grey matter cell layers in intralaminar and inter-regional neuronal communication (Parr et al., 2022).

Recall the scene between Mr. A and his husband's suggestion to try contemplative practices. At a basic sensory level, there are visual and auditory inputs that are explained at several levels in unimodal sensory cortex to determine attributes like line-orientation, luminance, shape, pitch, volume, and so on. As one progresses up the hierarchy in unimodal cortex, perceptual objects are extracted from the ambiguous stream of input: lips, mouth, face, phonemes, words, sentences. Higher still (in heteromodal association cortex), more complex predictions are employed: identifying the speaker as "husband," determining that the visual input of the moving mouth corresponds with the auditory input, forming semantic understanding, and so on.

At a higher level (involving heteromodal and amodal association cortex), Mr. A's husband's recommendation—do contemplative practice—is inferred as his husband giving unwanted advice. There is an urge to respond with anger (which would involve its own predictive cascade of action sequences), but another prediction interferes—something like *if I respond with frustration, I will be perceived as "needy," and my husband will be angry*. With that prediction (of the consequences of the action-sequence that cascades from anger; cf. expected free energy) Mr. A instead elects an alternative prediction: *if I say nothing and agree, then my husband will not be angry and attack me*. This prediction guides the motor sequence corresponding to saying nothing—spanning higher-order amodal regions (the concept of *say nothing*) down to unimodal motor execution of oral muscle contraction.

Recall that interoception, just as exteroception, is a sensory state of the brain's Markov blanket. Crucially, in the Free Energy Principle, *the body is external to the brain*. The brain is not in direct contact with bodily states. Bodily states are also hidden states. These must be inferred just as much as states of the external world.

Homeostasis—conceptually equivalent to the self-organizing system remaining within its viable bounds—is therefore a predictive process. Specifically, homeostatic categories—such as levels of blood oxygenation, sodium, water, temperature, glucose, blood pressure, and so on—are "hyperprior predictions." They dictate the viable bounds in which the system must remain to continue to exist. Hyperpriors predictions are distinct from predictions formed about the external world insofar as they *must* be confirmed. For example, the homeostatic hyperprior of viable internal temperature *must* be verified, or else the organism ceases to exist. Hyperpriors are innate predictions that cannot be modified or updated. Prediction errors related to hyperpriors cannot be minimized via perception (i.e., belief updating). One cannot update one's beliefs about viable glucose ranges without eventual starvation. Prediction errors deriving from hyperpriors can only be resolved via active blanket states (more complexities regarding mammalian homeostasis are discussed in Chap. 5).

Consider hunger. Hypothalamic nuclei detect changes in blood glucose levels. These interoceptive sensory inputs are compared to the hyperprior dictating the viable bounds of blood glucose levels. The comparison registers prediction error. When the prediction error reaches a high enough value (i.e., the deviation from the hyperprior prediction is too great; variational free energy increases) a predictive cascade is executed to return the blood glucose level within viable bounds (i.e., minimize the prediction error). This can involve a variety of action-predictions: releasing fat stores, consuming food, and so on. Such an action-cascade is driven by the prediction that these actions will result in sensory states that confirm the hyperprior prediction, returning blood glucose to the appropriate level.

Precision

Applying the free energy principle to human experience illustrates a key problem faced by complex self-organizing systems. Predictions and prediction errors range multiple levels and categories, some requiring more predictive work than others depending on (beliefs about) the current context. How do self-organizing systems determine which prediction errors can best drive model updating and action selection to minimize free energy? Conversely, how do self-organizing systems determine which predictions to employ to resolve prediction errors? Recall Mr. A's strategy of "separating" tasks to optimize outcomes. Why was this strategy (as opposed to, say, a holistic approach) Mr. A's mode for dealing with work-related problems? Or, why was Mr. A's husband's comment to "do contemplative practice" inferred as unwanted advice rather than an attempt at connection?

Here enters the concept of "precision." Precision refers to the confidence or salience of a prediction or prediction error. With respect to predictions as probability distributions, precision is the spread of the distribution. The precision of a prediction quantifies the likelihood of that prediction reducing prediction error. Very precise predictions hold high confidence—the prediction that a "face" is the cause of a specific constellation of sensory inputs (two ovals above a horizontal, oblong shape) is very precise—hence humans are prone to "seeing faces" in many scenes. The individual fruits and vegetables in Giuseppe Arcimboldo's Rudolf II of Habsurg as *Vertumnus* (1590) are individual objects, but the sensory input is grouped into a higher gestalt in humans whose perception of faces is very precise. A "face" is a precise explanation of this set of sensory inputs, even though it consists of different fruits and vegetables.

Precise prediction errors have high salience, meaning they are given more weight to drive predictive work. For instance, homeostatic hyperpriors (very precise predictions; they *must* be confirmed, and their confirmation minimizes prediction error by definition) often generate salient prediction errors. When blood oxygenation levels are low, very precise error signals register the difference between this sensory input and the hyperprior. This "suffocation alarm" drives actions to quickly remedy to situation (desperately moving in the direction of greater oxygen, gasping for breath, etc.)—that is, to minimize the free energy indexed to the hyperprior for blood oxygenation. In contrast, one's "beliefs" of what one is currently seeing are incredibly imprecise. The brain easily updates inferences regarding what is seen when visual sensory inputs change. Simply put, when you move your eyes to a new location, you "see" new objects (not the objects you were seeing a moment ago).

In the brain, precision is implemented through neuromodulatory control systems such as dopamine, acetylcholine, serotonin, norepinephrine, histamine, substance P, testosterone, estrogen, progesterone, and so on. Neuromodulatory control differs from classical neuronal transmission (i.e., synaptic communication via the release of neurotransmitters) in their greater spatial and temporal reach (i.e., they impact a greater range of brain areas at a time, and last longer). Neuromodulators control the post-synaptic gain of neurons—effectively modulating the weight of activity achieved by synaptic communication (see Solms, 2021 for details). These different neuromodulators have been associated with controlling precision of different classes of predictions and prediction errors such as interoception, exteroception, attentional gain, policies, and so on (see Parr et al., 2022).

Altogether, the generative model involves a set of hierarchical predictions with differing precision-weights. Prediction errors with high precision are more likely to be passed up the hierarchy to drive model updating and action-execution. Likewise, predictions with high precision are more likely to be employed. Less precise signals are attenuated, and less precise predictions are not selected. Self-organizing systems can thereby minimize prediction errors in three ways: model updating of prior predictions (perception), changing inputs (action), or adjusting precision-weights afforded to error signals.

In the scene of Mr. A and his husband, the prediction "unwanted advice" can be considered the most precise prediction to explain the sensory input of "do yoga"—more precise than, say, an alternative inference of "trying to connect." Moreover, the two conflicting predictions that follow—"respond with anger" and "say nothing"—differ not only in their inferred consequences but also in their precision. "Say nothing" is inferred to be more precise—that is, it is inferred to result in less prediction error than "respond with anger." Because of its higher precision-weight, that predictive cascade ("do nothing") drives the action sequence.

This discussion reveals a crucial point regarding precision. Mr. A's choice to "say nothing" is *inferred* to result in less prediction error. In other words, *precision itself must be predicted* (Friston, 2021). Precision-weights are changed based on predictions of how well a prediction can minimize free energy. This opens a key perspective on what changes might occur in therapy—such as problematizing (the precision of) the prediction "do nothing" and perhaps increasing the precision of a predictive cascade that could involve greater emotional expression.

This quasi-technical discussion of the Free Energy Principle provides a conceptual basis for a key cornerstone of Solms's neuropsychoanalytic model. My hope is that the abstract concepts in this chapter will become clearer through their grounding in Freudian terms. I now turn to this neuropsychoanalytic model.

References

Arcimboldo, G. (1590). *Vertumnus* (oil on canvas). Skokloster Castle.
Friston, K. (2010). The free-energy principle: A unified brain theory? *Nature, 11*(2), 127–138. https://doi.org/10.1038/nrn2787
Friston, K. (2021). Consciousness and felt uncertainty: Commentary on the hidden spring: A journey to the source of consciousness. *Journal of Consciousness Studies, 28*(11–12), 178–189. https://doi.org/10.53765/20512201.28.11.178
Hodson, R., Mehta, M., & Smith, R. (2024). The empirical status of predictive coding and active inference. *Neuroscience & Biobehavioral Reviews, 157*, 105473. https://doi.org/10.1016/j.neubiorev.2023.105473
Kaplan-Solms, K., & Solms, M. (2002). *Clinical studies in neuro-psychoanalysis: Introduction to a depth neuropsychology* (2nd ed.). Karnac Books.
Parr, T., Pazzulo, G., & Friston, K. (2022). *Active inference: The free energy principle in mind, brain, and behavior*. MIT Press.
Solms, M. (2021). *The hidden spring: A journey to the source of consciousness*. Profile Books.
Solms, M., & Turnbull, O. (2002). *The brain and the inner world: An introduction to the neuroscience of subjective experience*. Other Press.

6

Mark Solms's Neuropsychoanalytic Meta-Neuropsychology

Abstract This chapter outlines Mark Solms's meta-neuropsychology. I synthesize his arguments for dynamically localizing the ego to the neocortex and the id to subcortical, upper brainstem regions. By integrating Jaak Panksepp's affective neuroscience, Solms makes the profound argument that the id is affectively consciousness. One can then dynamically localize Freud's systems conscious, preconscious, and unconscious to functional neuroanatomy. I additionally discuss how Solms integrates Friston's Free Energy Principle into his meta-neuropsychology of a predictive ego seeking to minimize uncertainty rendered salient by the affective id.

Keywords Ego • Id • Affective consciousness • Panksepp • Neuroscience • Freud • Friston • Free energy • Prediction • Repression

Mark Solms coined the term neuropsychoanalysis and solidified the discipline as a bridge between psychoanalysis and the neurosciences. The

---

Here I revisit and expand discussions of Solms's oeuvre began in Dall'Aglio (2019, 2021a, 2022).

© The Author(s), under exclusive license to Springer Nature Switzerland AG 2024
J. Dall'Aglio, *A Lacanian Neuropsychoanalysis*, The Palgrave Lacan Series, https://doi.org/10.1007/978-3-031-68831-7_6

key methodological innovation was the use of the clinico-anatomical neurological method from a psychoanalytic lens (Kaplan-Solms & Solms, 2002). This refers to the study of patients with focal brain injury to draw connections between brain areas and mental functions. The clinico-anatomical method was a foundational approach in behavioral neurology. Moreover, Freud himself employed this approach in his neurological work (Freud, 1891; Solms & Saling, 1986).

Psychodynamic clinical work with patients with focal brain injury allows one to observe changes to the mental apparatus (in psychoanalytic terms) associated with a specific brain area. By collecting similar cases, common changes across patients can be used to infer the metapsychological operation impaired by the lesion. Importantly, Solms emphasizes an approach of "dynamic localization" (see Chap. 3) in the tradition of the Russian neuropsychologist (who also studied psychoanalysis) Alexander Luria. To briefly recount the principle of dynamic localization, a single brain area is involved in multiple psychical operations, and a psychical operation involves multiple brain areas. One therefore does not expect a one-to-one mapping, but broad links at the level of networks and larger brain regions. Neuropsychoanalysis admits of a broad localization without falling into the reductionist fallacy of a psychical process housed *inside* a neural center. Psychical functions emerge in the constellation *between* multiple neural centers.

The clinico-anatomical method and the lens of dynamic localization have been crucial theoretical and methodological principles for a psychoanalytic mapping of the brain. Notably, this approach allows concepts from different psychoanalytic schools to be mapped (Dall'Aglio, 2019; Morin, 2018; Salas et al., 2021). Furthermore, it allows one to (re)consider psychoanalytic ideas in the light of modern neuroscientific understanding of the brain (cf. threefold movement; see Chap. 3).

Such a sentiment is liable to raise eyebrows of those doubtful over whether neuroscience can contribute to psychoanalysis (as discussed in Chap. 2). Here, one should recall Freud's ongoing position that, despite the limitations of the biology of his time, there would come a day when biology could contribute to psychoanalytic meta-psychology. Such reflections abound throughout Freud's oeuvre; his musings in *Beyond the Pleasure Principle* are but one example:

On the other hand it should be made quite clear that the uncertainty of our speculation has been greatly increased by the necessity for borrowing from the science of biology. Biology is truly a land of unlimited possibilities. We may expect it to give us the most surprising information and we cannot guess what answers it will return in a few dozen years to the questions we have put to it. They may be of a kind that will blow away our whole artificial structure of hypotheses. (Freud, 1920, p. 60)

In a sense, neuropsychoanalysis has a less ambitious hope for biology than Freud here. In my view, neuroscience cannot "disprove" psychoanalysis in a straightforward fashion (cf. threefold movement). Rather, possibilities suggested through dialogue with neuroscience can lead psychoanalysis to reconsider its concepts and clinical applications and, from that basis, explore what changes and possibilities emerge. Solms's work illustrates precisely this.

A Psychoanalytic Mapping of the Brain

The Cortical Ego

Synthesizing literature on the functions of the neocortex and through psychodynamic work with neurologic patients with cortical lesions, Solms dynamically localizes the Freudian ego to the neocortex (Kaplan-Solms & Solms, 2002; Solms, 2013). Various functions that Freud (1923) attributed to the ego are performed by the human neocortex. These include perception, action, thinking, decision-making, inhibition, temporal ordering, and an orientation toward external reality. Posterior neocortex (especially heteromodal association cortex) generates representations of the external world. Solms calls these representations "mental solids." They are mental objects (re-representations of sensory experience) that can be held in mind through working memory (dynamically localized to prefrontal regions). Cortical representation gives stability to perceptions, and cognitive control capacities allow the organism to slow down (i.e., inhibit prepotent habitual responses), plan, and think (i.e., imagine action-sequences) before acting in the world.

Cortical processes are largely declarative—that is, they can be held before self-reflective consciousness and spoken about. Declarative systems include semantic and episodic memories. They are dependent on the hippocampal formation for encoding but are believed to be gradually transferred into long-term neocortical regions through systems consolidation (Goto, 2022; Takehara-Nishiuchi, 2021; Wiltgen & Tanaka, 2013) Episodic memories are ego-centric memories—for example, of what you ate for dinner yesterday. You can generate an image—a mental solid, with visual, auditory, tactile, olfactory, and gustatory qualities—of this memory and think about it, manipulate it in mind, and so on. Semantic memories are generalized facts about the world, such as the name of your aunt or the capital of France (Squire, 2004). These memories share the quality of being declarable, spoken about, and held in working memory. In other words, we can *think* about (and with) these memories.

The Subcortical Id

In contrast, functions that Freud (1923) attributed to the id can be dynamically localized to the upper brainstem and limbic structures involved in affect and motivation—especially the ascending reticular activating system (Solms, 2013). These regions are subcortical (deep grey matter areas, in contrast to the neocortex which sits at the periphery of the brain). Recall that the Freudian id includes the drives as representatives of the internal milieu of the body. Various diencephalic (e.g., hypothalamus, pituitary gland) and brainstem (e.g., pons, medulla) structures are especially sensitive to the internal body, detecting levels of blood oxygenation, heart rate, blood pressure, glucose, and so on.

Based on Panksepp's (Panksepp, 1998) cross-mammalian research on basic emotional systems (primarily using deep brain stimulation and pharmacological methods), Solms (2021a) revises Freudian drive theory to include seven emotional instincts. Just like bodily need-detectors, each involves a certain homeostatic set-point, viable bounds in which the organism can exist (see Table 6.1). However, these emotional instincts differ from bodily need-detectors insofar as their set-points are reduce the uncertainty generated by the deviation from the temperature set-point (more human-specific examples below, including the issue of the repressed unconscious).

Levels of Predictions

The ego thereby involves a range of predictions that can be charted according to the brain's different memory systems (Solms, 2021b; Squire, 2004). Some predictions are innate hyperpriors, such as body temperature and—importantly—the socio-emotional set-points of Panksepp's emotional systems (I return to this below). Other predictions are learned.

When applying the Free Energy Principle to the brain, one must therefore recognize a *predictive hierarchy*. In other words, there are *levels of predictions*. Solms (2021b) describes this hierarchy as organized like an onion. The deepest layers—innate hyperpriors and instincts—consist of predictions with high precision, high (temporal and spatial) generalizability, and low plasticity. In other words, they are applied across many situations and have high confidence that they will reduce prediction error. The hyperprior to remain within certain blood oxygenation bounds is always true for the organism and is sure to minimize uncertainty in this domain. Furthermore, uncertainty is minimally tolerated at this deep level. Deviations from these set-points are life-threatening; hence the errors are weighted with high precision and quickly drive predictive cascades to minimize uncertainty (e.g., suffocation alarm). Neuropsychologically, one can add the insight that such deep predictions are non-declarative.

As one moves toward the periphery of the onion-hierarchy, there is increasing tolerance of uncertainty, precision decreases, predictions become less generalizable (more specific and more complex), and plasticity increases. Neuropsychologically speaking, when these predictions reach semantic, episodic, and short-term memory systems, they are declarative. For example, I went to a local café that typically closes at 7:00pm but saw a sign that the café would close at 6:00pm for a staff meeting. The prior prediction *closes at 7:00pm* was compared to the sensory input *closes at 6:00pm*, the error signal is registered, and the prior prediction is updated. This updating was minimally problematic (despite my chagrin), did not sustain attention, and was easily dealt with. Uncertainty in the café environment—how many people sat at chairs, what music was playing, light levels, the décor, and so on—was highly complex yet comparably unimportant; that is, this information was weighted with low precision because it pertained to the periphery of the predictive hierarchy and was irrelevant to the writing-related goals I was attending to. Error signals here were unlikely to reach deeper layers, weighted with lower precision (see below), and thus uncertainty was tolerated and not passed up the hierarchy. Hence I have no memory of any perceptual details from that afternoon, nor did I notice most changes in the environment when sitting there.

In sum, predictive cascades involve a range of predictions at different levels. Homeostatic set-points drive innate reflexes that may be supplemented by other procedural memories and guided by semantic and episodic information to tailor action-sequences to the specific environment. Thinking further enables the selection of action-sequences with the lowest expected free energy.

The Repressed as Prematurely Automatized Predictions

However, this computational presentation should not be mistaken for seamlessness within the hierarchy. Conflicts between levels of predictions are readily apparent in clinical phenomena. Consider Mr. A. Awareness of the fact that he "sabotages himself" (involving declarative predictions) does not change the repetition of habitual actions (non-declarative predictions) that effect this self-sabotage. Here enters Solms's dynamic unconscious. For Solms, the repressed (non-declarative action plans that do not work) consists of *prematurely automatized predictions*. These are non-declarative predictions (action plans) weighted with high precision despite the fact that they are poor at minimizing free energy. They are resistant to updating (i.e., their precision-weights cannot be changed) given their low plasticity, and they have high generalizability insofar as they are applied in a range of situations. These features derive from the high precision afforded to these predictions at a deep level of the hierarchy.

As an aside, this is another instance of "weak nature" (Johnston, 2019; see Chap. 3). Very precise predictions can be laid down even though they are poor at minimizing free energy. Although the brain operates according to the Free Energy Principle, the complexity of the system leads to faults and short-circuits such as prematurely automatized predictions. (I continue this discussion in Chap. 8.)

Affective Consciousness as Precision-Weight Modulation: The Prioritization Function

In this view, Panksepp's emotional systems are hyperpriors at the deepest levels of the predictive hierarchy. These endogenous socio-emotional homeostatic set-points are precise predictions of the preferred states for the organism to occupy (e.g., stay close to caregivers; stay safe from bodily injury). At this level, deviations from these hyperpriors generate feelings of unpleasure. Within the Free Energy Principle, feelings therefore index a particular category of prediction error. For Solms, affective consciousness is *felt uncertainty*—the felt registration of deviations that correspond with increasing (informatic) uncertainty. This is why emotional homeostasis registers the state of the subject (as opposed to bodily need-detectors registering the state of the body); feelings indicate how well the predictive subject (i.e., the self-organizing system) is doing with respect to these endogenous set-points.

However, affects are not simply error signals, although they indicate deviations from hyperprior set-points. More specifically, affective consciousness adjusts the precision-weights throughout the predictive hierarchy (via neuromodulatory adjustment of post-synaptic gain). It does so by *palpating* throughout the hierarchy—*feeling one's way through different predictions* to adjust their precision based on how well they move the system toward these set-points. Predictions which decrease uncertainty (i.e., feel pleasurable) have their precision-weights increased; predictions which increase uncertainty (i.e., feel unpleasurable) have their precision weights decreased. Here, feeling is the guide, insofar as pleasure guides the adjustment of precision. Affective consciousness guides (predictions of) precision of predictions throughout the hierarchy.

For Solms, this is the function of consciousness. He notes that the critical mechanistic features of consciousness include:

(1) the presence in the system of *multiple* categories of demand for homeostatic work, which (2) need to be *prioritized* in a context-dependent manner, including (3) in *unpredicted* contexts where (4) *choices* must be made, and therefore (5) *voluntary* actions must be executed. (Solms, 2020b, p. 104, emphases in the original)

Several consequences follow. As stated, deviations from these set-points generate particular *qualities* of pleasure-unpleasure. At the mammalian level of subcortical brainstem consciousness, several *categories* of prediction errors (as feelings) are introduced. Panksepp's systems are encoded as categorical *qualitative* variables that cannot be reduced into a single, common *quantitative* factor. For instance, 3/5 SEEKING cannot be converted into 5/8 RAGE. One cannot resolve FEAR-uncertainty by eating. In other words, Panksepp's systems index different *categories of uncertainty*. Affect *guides* the organism according to a certain quality of uncertainty. Reduction of PANIC-uncertainty is the *feeling* of being close and secure with caregiving objects, which privileges different predictions than RAGE. Affective consciousness is *feeling one's way through uncertainty.*

The link between affective consciousness and uncertainty underscores that affective consciousness arises *in situations of uncertainty*—that is, in situations where the prediction(s) to resolve uncertainty are unclear. Hence the necessity of choice and voluntary action. In certain contexts, where the predictions to eliminate uncertainty are clear, consciousness is not needed. Indeed, most of our cognitive operations go on without any consciousness (Bargh & Chartrand, 1999); they minimize uncertainty without any need for conscious feeling. Likewise, we are not conscious of autonomic control of, for instance, blood pressure. It is in situations where the predictive cascade is *uncertain* that consciousness is necessary.

This introduces a crucial function to affective consciousness which I call the "prioritization function." Following Merker (2007) and Panksepp (1998), Solms (2020a) dynamically localizes this function to the midbrain decision triangle: the periaqueductal grey (the common terminal point for all of Panksepp's emotional systems), the midbrain tectum (registering a simplified saliency map of the external world), and the final motor output pathway. By integrating information about the state of the system (affect) and the environment (i.e., prior predictions regarding the present context), the prioritization function of affective consciousness selects which emotional system has the best likelihood of minimizing free energy. That system is *prioritized* and palpates through the predictive hierarchy, adjusting precision-weights according to that particular category of uncertainty (i.e., that particular category of *feeling*). Neurobiologically, this corresponds to adjusting post-synaptic gain through command neuromodulators associated with Panksepp's emotional systems. Psychologically, this corresponds to feeling.

Recall Mr. A's fright at the snake. In this situation, environmental and affective information was integrated: glucose-levels, temperature, proximity to caregivers (PANIC), the need to explore novelty (SEEKING), obstacles to goals (RAGE), bodily safety (FEAR), and so on. These are all potential categories of uncertainty whose palpations would render different predictive cascades more precise (e.g., SEEKING would lead to approach, engagement, and curious exploration; RAGE would lead to frustrated attack). In this situation, FEAR was inferred (recall that precision must be predicted) as the most salient category of uncertainty. In informatic terms, FEAR was prioritized. In neurobiological terms, neuromodulators involved in FEAR adjusted post-synaptic gain throughout the brain. In psychological terms, Mr. A *felt terrified*. FEAR prioritization adjusted precision-weights to employ predictive cascades that best minimized uncertainty in the domain of this socio-emotional homeostat. The man's advice to "separate" became a learned prediction, weighted with high precision, to minimize FEAR-uncertainty.

Additionally, recall Mr. A's conflict between frustration and upsetting his husband. One might cast this as the difference between RAGE and FEAR. Mr. A inhibited speaking from a place of anger and stayed silent, a form of avoidance. That is, FEAR was prioritized and guided Mr. A's predictive hierarchy. As a momentary aside, here one can ask: *why* was FEAR prioritized instead of RAGE? Why was FEAR *predicted* to be the category with greater likelihood of uncertainty reduction (i.e., higher precision) instead of RAGE? (I will return to this in Chap. 12.)

The operations of the prioritization function point to the idiosyncrasies of affective consciousness and the lack of a straightforward socio-emotional homeostasis. I propose that a Lacanian perspective—specifically that of *jouissance* and Lacanian register theory (see Chap. 4)—allows one to extract the radical dimensions of Solms's model. It is to this Lacanian neuropsychoanalytic integration that I now turn.

References

Bargh, J., & Chartrand, T. (1999). The unbearable automaticity of being. *American Psychologist, 54*(7), 462–479. https://doi.org/10.1037/0003-066X.54.7.462

Bazan, A. (2023). Primary and secondary process mentation: Two modes of acting and thinking from Freud to modern neurosciences. *Neuropsychoanalysis.* https://doi.org/10.1080/15294145.2023.2284697

Bazan, A., & Detandt, S. (2015). Trauma and jouissance: A neuropsychoanalytic perspective. *Journal of the Center for Freudian Analysis and Research, 26,* 99–127. https://jcfar.org.uk/wp-content/uploads/woocommerce_uploads/2015/12/JCFAR-26-Ariane-Bazan-and-Sandrine-Detandt.pdf

Bloomstedt, P., Hariz, M., Lees, A., Silberstein, P., Limousin, P., Yelnik, J., & Agid, Y. (2008). Acute severe depression induced by intraoperative stimulation of the substantia nigra: A case report. *Parkinsonism & Related Disorders, 14*(3), 253–256. https://doi.org/10.1016/j.parkreldis.2007.04.005

Claparède, E. (1911/1951). Reconnaissance et moitié [Recognition and "meness"]. *Archives de Psychologie, 11,* 79–90.

Dall'Aglio, J. (2019). Of brains and Borromean knots: A Lacanian meta-neuro psychology. *Neuropsychoanalysis, 21*(1), 23–38. https://doi.org/10.1080/15294145.2019.1619091

Dall'Aglio, J. (2021a). Sex and prediction error, part 2: Jouissance and the free energy principle in neuropsychoanalysis. *Journal of the American Psychoanalytic Association, 69*(4), 715–741. https://doi.org/10.1177/00030651211042377

Dall'Aglio, J. (2021b). What can psychoanalysis learn from neuroscience? A theoretical basis for the emergence of a neuropsychoanalytic model. *Contemporary Psychoanalysis, 57*(1), 125–145. https://doi.org/10.1080/00107530.2021.1894542

Dall'Aglio, J. (2022). In G. Gargiulo & J. Turtz (Eds.), *Neuropsychoanalysis: What, how, and why. In Enriching psychoanalysis: Integrating concepts from contemporary science and philosophy* (pp. 119–146). Routledge. https://doi.org/10.4324/9781003271499-11

Freud, S. (1891). *On aphasia: A critical study* (E. Stengel, Trans.). International Universities Press.

Freud, S. (1915/1957). The unconscious. In *The standard edition of the complete psychological works of Sigmund Freud, Vol. XIV* (J. Strachey, Ed., Trans.) (pp. 159–215). Hogarth Press.

Freud, S. (1920/1955). Beyond the pleasure principle. In *The standard edition of the complete psychological works of Sigmund Freud, Vol. XVIII* (J. Strachey, Ed., Trans.) (pp. 1–64). Hogarth Press.

Freud, S. (1923/1961). The ego and the id. In *The standard edition of the complete psychological works of Sigmund Freud, Vol. XIX* (J. Strachey, Ed., Trans.) (pp. 1–66). Hogarth Press.

Friston, K. (2010). The free-energy principle: A unified brain theory? *Nature, 11*(2), 127–138. https://doi.org/10.1038/nrn2787

Friston, K., & Stephan, K. (2007). Free-energy and the brain. *Synthese, 159*(3), 417–458. https://doi.org/10.1007/s11229-007-9237-y

Goto, A. (2022). Synaptic plasticity during systems memory consolidation. *Neuroscience Research, 183*, 1–6. https://doi.org/10.1016/j.neures.2022.05.008

Johnston, A. (2019). *Prolegomena to any future materialism, Volume two: A weak nature alone*. Northwestern University Press.

Kaplan-Solms, K., & Solms, M. (2002). *Clinical studies in neuro-psychoanalysis: Introduction to a depth neuropsychology* (2nd ed.). Karnac Books.

McIntyre, C., McGaugh, J., & Williams, C. (2012). Interacting brain systems modulate memory consolidation. *Neuroscience & Biobehavioral Reviews, 36*(7), 1750–1762. https://doi.org/10.1016/j.neubiorev.2011.11.001

Merker, B. (2007). Consciousness without a cerebral cortex: A challenge for neuroscience and medicine. *Behavioral and Brain Sciences, 30*(1), 63–81. https://doi.org/10.1017/S0140525X07000891

Morin, C. (2018). *Stroke, body image, and self-representation: Psychoanalytic and neurological perspectives* (K. Valendinova & C. Morin, Trans.). Routledge.

Moruzzi, G., & Magoun, H. (1949). Brain stem reticular formation and activation of the EEG. *Electrocephalography and Clinical Neurophysiology, 1*, 455–473. https://doi.org/10.1016/0013-4694(49)90219-9

Panksepp, J. (1998). *Affective neuroscience: The foundations of human and animal emotions*. Oxford University Press.

Parvizi, J., & Damasio, A. (2003). Neuroanatomical correlates of brianstem coma. *Brain, 126*(7), 1524–1536. https://doi.org/10.1093/brain/awg166

Penfield, W., & Jasper, H. (1954). *Epilepsy and the functional anatomy of the human brain*. Little & Brown.

Salas, C., Turnbull, O., & Solms, M. (Eds.). (2021). *Clinical studies in neurospychoanalysis revisited*. Routledge.

Shannon, C. (1948). A mathematical theory of communication. *Bell System Technical Journal, 27*(3), 379–423. https://doi.org/10.1002/j.1538-7305.1948.tb01338.x

Snider, S., Hsu, J., Darby, R., Cooke, D., Fischer, D., Cohen, A., Grafman, J., & Fox, M. (2020). Cortical lesions causing loss of consciousness are anticorrelated with the dorsal brainstem. *Human Brain Mapping, 41*(6), 1520–1531. https://doi.org/10.1002/hbm.24892

Solms, M. (2013). The conscious id. *Neuropsychoanalysis, 15*(1), 5–19. https://doi.org/10.1080/15294145.2013.10773711

Solms, M. (2017). What is "the unconscious," and where is it located in the brain? A neuropsychoanalytic perspective. *Annals of the New York Academy of Sciences, 1406*(1), 90–97. https://doi.org/10.1111/nyas.13437

Solms, M. (2018). The neurobiological underpinnings of psychoanalytic theory and therapy. *Frontiers in Behavioral Neuroscience, 12*, 294. https://doi.org/10.3389/fnbeh.2018.00294

Solms, M. (2020a). New project for a scientific psychology: General scheme. *Neuropsychoanalysis, 22*(1–2), 5–35. https://doi.org/10.1080/15294145.2020.1833361

Solms, M. (2020b). Response to the commentaries on the "New Project". *Neuropsychoanalysis, 22*(1–2), 97–107. https://doi.org/10.1080/15294145.2020.1843215

Solms, M. (2021a). Revision to drive theory. *Journal of the American Psychoanalytic Association, 69*(6), 1033–1091. https://doi.org/10.1177/0003065121105

Solms, M. (2021b). *The hidden spring: A journey to the source of consciousness.* Profile Books.

Solms, M., & Saling, M. (1986). On psychoanalysis and neuroscience: Freud's attitude to the localizationist tradition. *The International Journal of Psychoanalysis, 67*(4), 397–416.

Squire, L. (2004). Memory systems of the brain: A brief history and current perspective. *Neurobiology of Learning and Memory, 82*(3), 171–177. https://doi.org/10.1016/j.nlm.2004.06.005

Takehara-Nishiuchi, K. (2021). Neurobiology of systems memory consolidation. *European Journal of Neuroscience, 54*(8), 6850–6863. https://doi.org/10.1111/ejn.14694

Wiltgen, B., & Tanaka, K. (2013). Systems consolidation and the content of memory. *Neurobiology of Learning and Memory, 106*, 365–371. https://doi.org/10.1016/j.nlm.2013.06.001

Zúñiga, J. F. M. (2017). EnRAGEd: Introductory notes on aggression in a case of orbitofrontal syndrome. *Neuropsychoanalysis, 19*(1), 77–86. https://doi.org/10.1080/15294145.2017.1295816

7

Jouissance is Surplus Prediction Error

Abstract Here, I develop my central argument: *jouissance* corresponds to surplus affective consciousness, to surplus (prioritized) free energy. I detail how Solms's neuropsychoanalytic integration of affective neuroscience and the free energy principle can be interpreted through Lacan's extimate topology of *jouissance*. I then discuss several implications from this Lacanian neuropsychoanalytic integration: differentiating drive (and *jouissance*) from instinct (and affect), Ariane Bazan's work situating the signifier as a motoric element, and articulating predictions as signifiers.

Keywords Drive • Instinct • Signifier • Affective neuroscience • Motor trace • Incentive sensitization • Free energy principle • Lacan • Sex • Panksepp

---

This chapter expands arguments from Dall'Aglio (2021).

© The Author(s), under exclusive license to Springer Nature Switzerland AG 2024
J. Dall'Aglio, *A Lacanian Neuropsychoanalysis*, The Palgrave Lacan Series,
https://Doi.org/10.1007/978-3-031-68831-7_7

Jouissance in the Lens of Affective Neuroscience and the Free Energy Principle

Now I come to the crux of my argument. I propose that the Lacanian concept of *jouissance* corresponds to affective consciousness, specifically affective consciousness as *surplus prediction error*. Recall that *jouissance* names an excess excitation beyond the binding capacities of the symbolic and imaginary registers. In that sense, *jouissance* is non-representational, a "bulge in the phenomenal field." It is the excitatory index of the real, the excess of negativity. It exceeds the bounds of the pleasure principle, naming an enjoyment in the increase of tension. On the side of the drive, *jouissance* is the drive-tension demanding work from the ego. It thus has a paradoxical place as both external to but nevertheless central to the operations of the ego. Lacan's term for this paradoxical place is *extimacy*—a neologism condensing "exterior," "external," "internal," and "intimate" (Lacan, 1959–1960; Miller, 1988). What is most intimate is also most alien.

Likewise, affective consciousness corresponds to (prioritized) categories of prediction errors. Prediction error is the remainder—the excess or surplus—outside predictive work, what is not fully accounted for by prediction. Affective consciousness arises at the point where prediction fails, the point of uncertainty within the predictive model. Insofar as the neuropsychoanalytic ego is a predictive ego, surplus prediction error (affective consciousness) is both outside the ego's managerial capacities (i.e., beyond what is explainable by egoic prediction) as well as the necessary impetus for prediction (i.e., affective consciousness as surplus prediction error drives predictive work; affective consciousness is the prerequisite basis for egoic cognitive consciousness). As an excess of uncertainty that excites the organism, demanding predictive work, the surplus prediction error of affective consciousness (*felt uncertainty*) corresponds to Lacanian *jouissance*.

Affective consciousness is therefore logically external yet fundamental to egoic prediction. This is significant: the core of consciousness (affective consciousness as surplus prediction error) is logically external to the ego, the "I," the sense of self. Freud's "great insult" to man was the insight that at the core of man is the unconscious, that the ego is not master in its own house. In a Lacanian lens, Solms levies another "great insult": the fount of consciousness is not the ego but the id, something radically exterior to the "I." The id is the fount of affective consciousness, a *fount of jouissance*. Insofar as affective consciousness is *jouissance*, the core of consciousness is paradoxically external to the ego. Affective consciousness is *extimate* to the conscious ego (Dall'Aglio, 2022).

Several implications follow from this connection between affective consciousness and *jouissance*—I will unpack these in the remaining chapters. However, one will note a tension in this connection. Namely, Solmsian-Pankseppian affective consciousness operates on the side of homeostasis, following Friston's Law to minimize free energy. Lacanian *jouissance* notoriously drives a hole in homeostasis. How do I reconcile this contradiction?

The Difference between Drive and Instinct

Affective consciousness consists of Panksepp's emotional instincts that seek to maintain homeostatic set-points, the reduction of tension (in Freudian terms) or of prediction error/free energy (in Solmsian-Fristonian terms). In Lacanian psychoanalysis, this view of affective consciousness follows the logic of homeostatic *instinct*. On the other hand, *jouissance* follows the logic of *drive*—specifically, the tendency to be caught in repetitive circuits and derive enjoyment from the repetition itself, the ongoing and sustained tension (Lacan, 1964).

Indeed, this distinction explains various clinical phenomena. Consider eating disorders. The instinctual logic of hunger would dictate that, upon eating enough to return to the homeostatic set-point for blood sugar, one should stop eating. However, in anorexia, the patient continues to not eat. In bulimia, the patient eats beyond the bounds of hunger-homeostasis.

For Lacan, this is because drive differs from instinct insofar as its object is not a particular object in the world, such as the prototypical milk for the infant at the mother's breast. Rather, the object of the drive is the *lost* object, *objet a*. Drive orients toward the negative excess in the representational, graspable world—impossible to attain but nevertheless providing some enjoyment in the very failure.

Therefore, the anorexic does not simply "not eat." The anorexic "eats the nothing" (Lacan, 1964, p. 104). The anorexic engages in the act of not eating as a way of demarcating the contours of the real, the perimeter of *objet a*, to derive *jouissance* that sticks to this precipice. Likewise, the bulimic eats excessively because no particular object of hunger fills the lack of the drive—excessive eating demarcates and shapes this lack in what can be grasped. Drive orients toward the real of *jouissance*.

Affect versus *Jouissance*: Drive as Instinct Hooked into the Logic of Excess

From the perspective of neuroscience, however, there is no clear separation of which systems would be "drives" and which would be "instincts." SEEKING perhaps comes closest, as a "goad without a goal" that proactively engages with uncertainty. Nevertheless, this imperative serves the goal of ultimately reducing prediction error. Recall its set-point: engage proactively with uncertainty to maximize the likelihood of reducing that uncertainty.

Nevertheless, Panksepp's instincts are not simply innate reflexes. As already discussed, they come with "holes" built into them to allow experience-dependent learning. Objects and aims of Panksepp's systems are flexible—corresponding with Freud's meta-psychology of drive (Freud, 1915). Yet, for some this remains insufficient and is merely instinctual homeostasis with an adaptive twist (Samuels, 2022).

However, consider how Solms (2020) characterizes eating disorders. Anorexia and bulimia are not aberrations (principally) of the hunger instinct but of Pankseppian emotional systems: PANIC, RAGE, FEAR, and so on. For Solms, eating disorders are problematics of emotional needs which are being insufficiently met. Solms's neuropsychoanalytic explanation points to the idea of one system being "hooked" into the logic of another. Hunger-related predictions arise to deal with unsatisfied emotional systems. Or, unsatisfied emotional systems interrupt the typical operation of hunger. If Panksepp's systems are instincts, they are very strange instincts to say the least (another instance of "weak nature"; see Chap. 3).

I propose that the Lacanian distinction between drive and instinct is a false dichotomy that often serves to segregate psychoanalysis from other fields. The capital-A Animal as perfectly in balance with its environment is a theoretical strawman—an instance of belief in a non-barred Other (Johnston, 2019). One must instead insist that the Animal or the Instinct does not exist (see Chap. 3; Zupančič, 2017). Rather than maintaining a separation between drive and instinct, I propose to inscribe drive *into* instinct. Drive names the tendency of *all instincts* toward excessive repetition beyond homeostatic bounds (see below), to be prone to aberrant hooking into predictive cascades where they are not "naturally" determined to be. *Drive is the aberration of instinct.* More specifically, the aberrated instinct is oriented toward *objet a*—hence the ongoing repetition and manifest (self-reflectively conscious) dissatisfaction, the strange enjoyment (*jouissance*) in the suffering of the symptom.

Correspondingly, one can draw an internal distinction between affect—emotion as typically considered—and *jouissance*. Affect would name the normal rise and fall of (prioritized) prediction error within the managerial capacities of the predictive ego. These are deviations that can satisfyingly be dealt with. *Jouissance*, in contrast, names the *surplus* prediction error outside predictive management. Specifically, *jouissance* arises at the point of instinct being hooked into the logic of excess, being oriented toward *objet a*.

The *Fort-Da* example illustrates this point of instinctual aberration. Freud (1920) recounts that his grandson, upon his mother's departure, played a game of throwing and pulling back a cotton-reel, exclaiming "ooo" (*Fort*, gone) and "aaa" (*Da*, here). Although Freud notes greater satisfaction upon the return *Da*, his nephew repeated the throw *Fort* more frequently. If one simply reads this situation as the child substituting the cotton-reel for his mother, a difficulty remains. Why does the child repeat the departure more frequently? Why repeat the whole sequence at all? A substitutive-homeostatic logic does not explain the repetition.

One can read this scene neuropsychoanalytically. Suppose that Freud's grandson experiences PANIC upon his mother's separation. Rather than follow the innate PANIC sequences (separation distress vocalization, SEEKING out the mother), Freud's grandson instead PLAYs a game of *Fort-Da*. However, he does not PLAY this game in the normal predictive dictates of PLAY either, as it does not involve a social partner. Rather, Freud's grandson *creates* in the space *between* PANIC and PLAY—and it is here that he derives enjoyment (*jouissance*) in repetition. The *creation* of a novel predictive cascade—specifically one that does not eliminate the tension but instead *metabolizes* it (sustains it in such a way that is enjoyed; see Chap. 14)—occurs in this slippage and aberrancy of emotional systems.

In a Lacanian neuropsychoanalytic perspective, the aberration of instinct is not merely due to contingent circumstance of life or external traumatic factors. All instincts are *structurally* prone to this aberration due to the structure of the predictive system. One such structural feature is the "non-rapport" between drive-need and motor-aim.

Splitting-off of the Motor Axis

Bazan and Detandt (2013) stress that, with vertebrate organisms, there is a "cut" between internal bodily needs and the actions the organism must take to meet those needs.[1] For Bazan and Detandt, any need arousal creates tension, to which the organism responds with motoric helplessness. When a contingent action occurs that satisfies the need (the experience of satisfaction; Freud, 1895; Van de Vijver et al., 2017), there must be some system to solder the link between action and need, installing a minimal history within the nervous system.

Here, Bazan and Detandt (2013) highlight the dopaminergic spike mechanism. Midbrain dopamine spikes project from the ventral tegmental area to the striatal nucleus accumbens (adjacent to the basal ganglia) upon the encounter with surprise. Because there is no inherent (i.e., prior

---

[1] Here, Bazan is critical of the linkage in Pankseppian affective neuroscience between emotional arousal and instinctual response.

predicted) link between need and specific action (Freud, 1895), the experience of satisfaction is also an experience of surprise. The dopamine spike "tags" the motor trace to form an associative complex between need arousal, perceptual traces, and motor action.

Dopaminergic spike tagging, however, introduces a new dimension. Bazan and Detandt adapt a key distinction in addictions literature between "wanting" and "liking" (Berridge, 2007; Berridge & Robinson, 2016; Robinson & Berridge, 1993) In addictions, wanting refers to the compulsive craving for the drug whereas liking refers to the consummatory pleasure. In Pankseppian terms, wanting corresponds to the dopaminergic SEEKING system—specifically SEEKING an object previously encountered that provided a surprising satisfaction, a wishful cathexis (Alcaro & Panksepp, 2011; Solms, 2021a). Liking corresponds to "hedonic hotspots" scattered throughout the forebrain and cortex and include opioid-mediated mechanisms (among other neuromodulatory systems).

Dopamine spike-mediated wanting (or SEEKING) tags the motor (and perceptual) trace with "incentive sensitization" (Berridge, 2007). Incentive sensitization refers to a charge or change to the trace (via various neuroplastic mechanisms; Berridge & Robinson, 2016) that makes its repetition enjoyable (albeit not necessarily consciously)—independent from need-arousal or need-satisfaction. The incentive sensitized motor trace tends toward repetition in a compulsory fashion, a repetition that itself is enjoyable and distinct from consummatory pleasure. SEEKING here diverges—along with the motor and perceptual traces affiliated with SEEKING-arousal (Alcaro & Panksepp, 2011)—from the original need. Dopaminergic spike-tagging effects a cut between homeostasis and repetition of the motor axis—an enjoyed repetition that Bazan and Detandt (2013) link to *jouissance*. Indeed, this fits with Solms's characterization of SEEKING activation: a particular quality of feeling good associated with engagement with (i.e., increasing) uncertainty. Such a feeling quality approaches the idea of enjoyment in the increase of tension associated with *jouissance*—specifically the repetition of actions (sustaining or increasing tension) separate from satisfaction-related tension-reduction.

I propose expanding Bazan et al.'s model to all of Panksepp's instincts. The specific neural implementation is a topic for future research. One possibility is that there are multiple mechanisms of sensitization other than dopamine spike tagging, such as central sensitization in nociceptive transmission (Woolf, 2011), glutaminergic excito-toxicity (Olloquequi et al., 2018), glutamate-related sensitization and dendritic plasticity in mesolimbic neurons (Berridge & Robinson, 2016), and wider-ranging sensitization via changes in gene transcription and expression in brain areas responsive to acute and longer-term stress (Dimitriadis, 2017; Post, 1992). Another is the fact that all of Panksepp's systems are *confluent* with SEEKING (Solms, 2021c). Insofar as Panksepp's systems require some stance toward the external world, SEEKING is a common road to engagement with the (novel) world where contingent encounters rouse other systems. Moreover, the dopamine spike mechanism—which is part of the SEEKING circuit—is involved in learning motor actions as well as policies for active predictive cascades for multiple homeostatic categories (Badre, 2020; Dall'Aglio, 2023; Parr et al., 2022).

Regardless of the mechanism(s), *any* action laid down to manage the tension aroused by an emotional system can split off in its own autonomy from that system. Its repetition and execution thereby invoke a dimension *supplementary* to homeostasis. This excess is *jouissance*, the repetition involved in the instinct's aberrant hooking around *objet a*. Of note, the common principle (whether through SEEKING confluence, difference modes of sensitization) comes into view through the Free Energy Principle: changes of precision via modulation of post-synaptic gain leading to greater activity and repetition (Parr et al., 2022).

Furthermore, I propose a consilience between Bazan et al.'s tagged motor traces and Solms's prematurely automatized predictions. Both operate at the level of the basal ganglia, the intersection of SEEKING (dopamine spike mechanisms) and motor learning. The dopamine spike would correspond to the assignment of high precision to a motoric prediction. Increasing precision would facilitate its repetition which, recall, is the repetition of an originally surprising encounter. The repetition retains some of the enjoyment in this surprise through its affiliation with SEEKING (i.e., incentive sensitization; wishful cathexis). Hence, Solms's premature automatized repressed does not only repeat because of its high precision and location deep within the predictive hierarchy. The premature automatized repressed repeats because its repetition is enjoyed. It is a means of *jouissance*.

The Signifier as a Motoric Element

Bazan (2011) further emphasizes the motor axis as independent, *barred* from a natural connection to bodily homeostasis. This parallels the Lacanian split between signifier and signified. Speech is fundamentally a motor act which is notoriously ambiguous. The phonetic sound-stream *Can certain people win?* includes the potential for *Can certain people* as well as *Cancer in people*. Linguistic ambiguity arises in the interplay of phonemes which must be cut, divided, and ordered in some way (in accordance with our predictions) to extract some meaning (signified).

For Bazan, the signifier is a motoric-phonemic element whose ambiguity has operative effects through its activation of associated traces—some associations following phonological lines (e.g., *Can certain people* and *Cancer in people*). Such associations are activated but barred from full execution due to the inhibition necessary to produce meaning. Inhibition is especially employed in cases of emotional significance (cf. Freud's forgetting of *Signorelli*; Bazan, 2011). Several studies have demonstrated the subliminal activation and effectiveness of the signifier, its phonological associations, and the connection between inhibition of phonological ambiguity and anxiety (Bazan et al., 2019; Bruxelmane et al., 2020; Olyff & Bazan, 2023; Thieffry et al., 2023).

However, an inhibited motor trace does not simply lay dormant. Following Jeannerod (1994), Bazan (2011) proposes that inhibited motor traces remain active with a baseline tension because their activation—specifically, the efference copy signal, the *prediction* of feedback based on motor output—is not attenuated by proprioceptive-sensory feedback signals. This is because the inhibition prevents sensory feedback; the execution does not occur. Motor tension gives rise to motor representation (Jeannerod, 1994), especially in the case of sensitized motor traces (Bazan, 2011; Bazan & Detandt, 2013).

Motor representations function as higher-order motor control mechanisms that influence motor output and organize perceptual and semantic experience. Here is the mechanism for the effectiveness of the signifier. Bazan (2011) gives the example of the signifier *rat* in Freud's Rat Man case as an instance of a motoric-phonemic signifier whose tension is not fully attenuated. It is worked over into a higher-order motor representation that organizes particular instantiations: Frau Hof*rat*, the issues with money (*Ratten*), the father's gambling (*Speilrat*), the infamous rat-torture, and so on. Thus, Bazan's operative signifiers are motoric-phonemic traces whose tension cannot be fully eliminated. This might be due to inhibition, motoric immaturity that prevents full execution, or the ongoing incentive sensitized (SEEKING) activation. Or, as I develop in Chap. 8, the non-attenuated excitation of key motor elements could be due to the impossibility of fully eliminating the surplus prediction error of affective consciousness.

Predictions as Signifiers

A connection therefore appears between the motor axis and the signifier. In computational terms, if surplus prediction error corresponds to *jouissance*, then it might be said that predictions correspond to signifiers. Predictions explain sensory input and allow some minimization of prediction error; signifiers bind excess excitation and allow some metabolization of *jouissance* (Lacan, 1959–1960).

Predictions also (largely) derive from the social order insofar as they are learned (Holmes & Nolte, 2019). Even predictions associated with bodily homeostasis originate in the relationship with the primary caregiver (Fotopoulou & Tsakiris, 2017), and Pankseppian hyperpriors are object relational insofar as they imply a certain type of object (Kernberg, 2022; Solms, 2021b). For Lacan, the social order—the Other (family, culture, etc.)—is the repository of signifiers (Lacan, 1953–1954, Lacan, 1964).

Recall the case of Mr. A. "Separate" can be considered a prediction that derives from the social order (the man who gives the advice) that does something to manage the FEAR-prediction error. In Lacanian terms, the signifier "separate" binds *jouissance*. Nevertheless, in a Bazanian fashion, this signifier takes on weight beyond its immediate context and becomes a super-ordinate representation that influences Mr. A's general approach to life. In other words, it repeats *beyond* this FEAR-situation, and this repetition procures enjoyment.

This example illustrates the difficulty with the notion that "predictions *explain* sensory input." What does *explain* mean here? Explain means something different for motor (non-declarative) predictions than it does for semantic (declarative) predictions. For the latter, *explain* approximates semantic understanding. For the former, *explain* assuages FEAR prediction error through motor and cognitive actions. Both are minimizations of free energy but in qualitatively distinct hierarchical levels. In the example of Mr. A, the signification of "separate" means something different in the snake-context from his self-inhibition in the context of marital strife. Here, the Lacanian distinction between signifier and signified—and between real, symbolic, and imaginary—comes into play. I will return to this in Chaps. 9, 12, and 13. For now, it suffices to grasp that predictions can operate like signifiers. The disjuncture between signifier (motoric prediction) and drive-tension (prediction error as demand for predictive work) is a site of *jouissance* as surplus prediction error (affective consciousness).

Recall that *jouissance* arises at the point of antagonism immanent to the symbolic (see Chap. 4). If *jouissance* corresponds to surplus prediction error and signifiers correspond to predictions, then it follows that surplus prediction (affective consciousness) would emerge at the point of antagonism *immanent to the predictive system*. I suggest that this antagonism reaches deeper than the splitting of the motor axis, down to the prioritization function of affective consciousness. It is to this structural antagonism that I now turn.

References

Alcaro, A., & Panksepp, J. (2011). The SEEKING mind: Primal neuro-affective substrates for appetitive incentive states and their pathological dynamics in addictions and depression. *Neuroscience & Biobehavioral Reviews, 35*(9), 1805–1820. https://doi.org/10.1016/j.neubiorev.2011.03.002

Badre, D. (2020). *On task: How our brain gets things done*. Princeton University Press.

Bazan, A. (2011). Phantoms in the voice: A neuropsychoanalytic hypothesis on the structure of the unconscious. *Neuropsychoanalysis, 13*(2), 161–176. https://doi.org/10.1080/15294145.2011.10773672

Bazan, A., & Detandt, S. (2013). On the physiology of jouissance: Interpreting the mesolimbic dopaminergic reward functions from a psychoanalytic perspective. *Frontiers in Human Neuroscience, 7,* 709.

Bazan, A., Kushwaha, R., Winer, E. S., Snodgrass, J. M., Brakel, L., & Shevrin, H. (2019). Phonological ambiguity detection outside of consciousness and its defensive avoidance. *Frontiers in Human Neuroscience, 13,* 77. https://doi.org/10.3389/fnhum.2019.00077

Berridge, K. (2007). The debate over dopamine's role in reward: The case for incentive salience. *Psychopharmacology, 191*(3), 391–431. https://doi.org/10.1007/s00213-006-0578-x

Berridge, K., & Robinson, T. (2016). Liking, wanting, and the incentive-sensitization theory of addiction. *American Psychologist, 71*(8), 670–679. https://doi.org/10.1037/amp0000059

Bruxelmane, J., Shin, J., Olyff, G., & Bazan, A. (2020). Eyes wide shut: Primary process opens up. *Frontiers in Psychology, 11,* 145. https://doi.org/10.3389/fpsyg.2020.00145

Dall'Aglio, J. (2021). Sex and prediction error, part 2: Jouissance and the free energy principle in neuropsychoanalysis. *Journal of the American Psychoanalytic Association, 69*(4), 715–741. https://doi.org/10.1177/00030651211042377

Dall'Aglio, J. (2022). In G. Gargiulo & J. Turtz (Eds.), *Neuropsychoanalysis: What, how, and why. In Enriching psychoanalysis: Integrating concepts from contemporary science and philosophy* (pp. 119–146). Routledge. https://doi.org/10.4324/9781003271499-11

Dall'Aglio, J. (2023). Extending the theory of premature automatization: The fantasy as an abstract rule in hierarchical cognitive control. *Neuropsychoanalysis, 25*(1), 27–42. https://doi.org/10.1080/15294145.2023.2183888

Dimitriadis, Y. (2017). The psychoanalytic concept of jouissance and the kindling hypothesis. *Frontiers in Psychology, 8,* 1593. https://doi.org/10.3389/fpsyg.2017.01593

Fotopoulou, A., & Tsakiris, M. (2017). Mentalizing homeostasis: The social origins of interoceptive inference. *Neuropsychoanalysis, 19*(1), 3–28. https://doi.org/10.1080/15294145.2017.1294031

Freud, S. (1895/1966). Project for a scientific psychology. In *The standard edition of the complete psychological works of Sigmund Freud, Vol. 1* (J. Strachey, Ed., Trans.) (pp. 281–391). Hogarth Press.

Freud, S. (1915/1957). Instincts and their vicissitudes. In *The standard edition of the complete psychological works of Sigmund Freud, Volume 14* (J. Strachey, Ed., Trans.) (pp. 109–140). Hogarth Press.
Freud, S. (1920/1955). Beyond the pleasure principle. In *The standard edition of the complete psychological works of Sigmund Freud, Vol. XVIII* (J. Strachey, Ed., Trans.) (pp. 1–64). Hogarth Press.
Holmes, J., & Nolte, T. (2019). "Surprise" and the Bayesian brain: Implications for psychotherapy theory and practice. *Frontiers in Psychology, 10*, 592. https://doi.org/10.3389/fpsyg.2019.00592
Jeannerod, M. (1994). The representing brain: Neural correlates of motor intention and imagery. *Behavioral and Brain Sciences, 17*, 187–245. https://doi.org/10.1017/S0140525X00034026
Johnston, A. (2019). *Prolegomena to any future materialism, Volume two: A weak nature alone.* Northwestern University Press.
Kernberg, O. (2022). Some implications of new developments in neurobiology for psychoanalytic object relations theory. *Neuropsychoanalysis, 24*(1), 3–12. https://doi.org/10.1080/15294145.2021.1995609
Lacan, J. (1953–1954/1991). *The seminar of Jacques Lacan, Book I: Freud's papers on technique* (J.-A. Miller, Ed.; J. Forrester, Trans.). Norton.
Lacan, J. (1959–1960/1992). *The seminar of Jacques Lacan, Book VII: The ethics of psychoanalysis* (J.-A. Miller, Ed., & D. Porter, Trans.) Norton.
Lacan, J. (1964/1978). *The seminar of Jacques Lacan, Book XI: The four fundamental concepts of psychoanalysis* (J.-A. Miller, Ed., A. Sheridan, Trans.). Norton.
Miller, J.-A. (1988). Extimité. *Prose Studies, 11*(3), 121–131. https://doi.org/10.1080/01440358808586354
Olloquequi, J., Cornejo-Córdova, E., Verdaguer, E., Soriano, F., Binvignat, O., Auladell, C., & Camins, A. (2018). Excitotoxicity in the pathogenesis of neurological and psychiatric disorders: Therapeutic implications. *Journal of Psychopharmacology, 32*(3), 265–275. https://doi.org/10.1177/0269881118754680
Olyff, G., & Bazan, A. (2023). People solve rebuses unwittingly—Both forward and backward: Empirical evidence for the mental effectiveness of the signifier. *Frontiers in Human Neuroscience, 16*, 965183. https://doi.org/10.3389/fnhum.2022.965183
Parr, T., Pazzulo, G., & Friston, K. (2022). *Active inference: The free energy principle in mind, brain, and behavior.* MIT Press.
Post, R. (1992). Transduction of psychosocial stress into the neurobiology of recurrent affective disorder. *American Journal of Psychiatry, 149*, 999–1010. https://doi.org/10.1176/ajp.149.8.999

Robinson, T., & Berridge, K. (1993). The neural basis of drug craving: An incentive-sensitization theory of addiction. *Brain Research Reviews, 18*(3), 247–291. https://doi.org/10.1016/0165-0173(93)90013-p

Samuels, R. (2022). *(Mis)understanding Freud with Lacan, Zizek, and neuroscience*. Palgrave.

Solms, M. (2020). Mark Solms teaches from lock-down [Online course]. Teachable. https://mark-solms-in-lock-down.teachable.com/

Solms, M. (2021a). Commentary on Dall'Aglio. *Journal of the American Psychoanalytic Association, 69*(4), 767–774. https://doi.org/10.1177/00030651211037632

Solms, M. (2021b). Response to Otto Kernberg. *Neuropsychoanalysis, 23*(2), 115–119. https://doi.org/10.1080/15294145.2021.1984284

Solms, M. (2021c). Revision to drive theory. *Journal of the American Psychoanalytic Association, 69*(6), 1033–1091. https://doi.org/10.1177/0003065121105

Thieffry, L., Olyff, G., Pioda, L., Detandt, S., & Bazan, A. (2023). Running away from phonological ambiguity, we stumble upon our words: Laboratory induced slips show differences between highly and lowly defensive people. *Frontiers in Human Neuroscience, 17*, 1033671. https://doi.org/10.3389/fnhum.2023.103367

Van de Vijver, G., Bazan, A., & Detandt, S. (2017). The mark, the Thing, and the object: On what commands repetition in Freud and Lacan. *Frontiers in Psychology, 8*, 2244. https://doi.org/10.3389/fpsyg.2017.02244

Woolf, C. (2011). Central sensitization: Implications for the diagnosis and treatment of pain. *Pain, 152*(3), S2–S15. https://doi.org/10.1016/j.pain.2010.09.030

Zupančič, A. (2017). *What IS sex?* MIT Press.

8

The Neuronal Real: Antagonism Immanent to the Brain

Abstract If *jouissance* arises from the point of antagonism within the symbolic, and I claim that *jouissance* corresponds to (prioritized) surplus affective consciousness, then is it possible to formulate antagonism within the brain? Here, I demonstrate how antagonism is not only immanent to the brain's inherited structure; it is also necessary for affective consciousness. Consciousness depends on antagonism (the real) immanent to a brain organized as a differential system (the symbolic). This allows me to situate the Lacanian split subject ($), *objet a*, and *das Ding*—as well as develop the notion of an emotional system operating in the logic of *jouissance*: J(E).

Keywords Prediction • Predictive coding • Free energy principle • Consciousness • Jouissance • Lacan • Panksepp • Affective neuroscience • Neuropsychoanalysis, Friston

---

Here I expand arguments from Dall'Aglio (2021).

© The Author(s), under exclusive license to Springer Nature Switzerland AG 2024
J. Dall'Aglio, *A Lacanian Neuropsychoanalysis*, The Palgrave Lacan Series,
https://doi.org/10.1007/978-3-031-68831-7_8

Innate Affective Hyperpriors Conflict

Recall Solms's "critical mechanistic features of consciousness":

> (1) the presence in the system of *multiple* categories of demand for homeostatic work, which (2) need to be *prioritized* in a context-dependent manner, including (3) in *unpredicted* contexts where (4) *choices* must be made, and therefore (5) *voluntary* actions must be executed. (Solms, 2020b, p. 104, emphases in the original)

As reviewed in Chap. 6, the "multiple categories of demand for homeostatic work" refer to (at least) Panksepp's seven emotional systems—qualitatively distinct domains of uncertainty. Points (2) and (3) indicate the radical dimension of Solms's model. That "unpredicted contexts" require "context-dependent" prioritization shows that *these innate emotional systems conflict* (Solms, 2021a). PANIC demands proximity to caregivers; SEEKING propels one to see what else is out there. RAGE and FEAR toward a bad object promote conflicting action plans. When one's caregiver is frustrating, how to manage the conflict between PANIC and RAGE? A PLAYful game of tag is fun until someone gets severely hurt and FEARs further harm.

In computational terms, Pankseppian hyperpriors conflict—they sit in an uncomfortable hodge-podge, a Frankenstein-like kludge. Our evolutionary inheritance is not a seamless, harmonious, single homeostatic system. There are homeosta*ses* whose dictates regarding different categories of uncertainty in no way line up with each other. Such conflict is not an effect of external circumstances. The very *structure* of affective consciousness is marred by conflicting hyperpriors.

There is No Meta-Prediction

Importantly, there is no inherited resolution-mechanism for conflicting hyperpriors. Solms discusses:

> It is conceivable that an extremely complex set of model algorithms could evolve to compute relative survival demands in all predictable situations, to

8 The Neuronal Real: Antagonism Immanent to the Brain

enable us *automatically* to prioritise actions on this basis. However, such complex models are extremely expensive, in every sense of the world. They are unwieldly, which means delay, which can be the difference between life and death; and they require lots of processing power, which means having to find more energy resources. Statisticians call the exponential increase in computational resources necessitated by a linear increase in model complexity the 'combinatorial explosion'. (Solms, 2021b, pp. 193–194, emphasis in the original)

Solms's reflection here could be condensed into an aphorism: *There is no meta-prediction*. I cast this aphorism in tune with Lacan's *There is no meta-language*. For Lacan, *There is no meta-language* refers to the fact that there is no beyond of spoken language that guarantees its significations, its structure, or even the fact that it can be meaningful (Johnston, 2005, 2019; Žižek, 2020b). It is similar to the notion *There is no Other of the Other* (see Chap. 3).

This is why signifiers sit in an indeterminate relationship with signifieds, one signifier capable of referring to multiple signifieds depending on the particularities of the signifying chain. Any linkage or quilting between signifier and signified is a contingent knotting. In other words, *There is no meta-language* affirms the real as a structural antagonism immanent to the symbolic—there is no external guarantee that perfectly resolves the antagonism of the symbolic (Zupančič, 2017).

There is no meta-prediction can be read similarly. Solms notes that the brain could not inherit any algorithm that could resolve all possible conflicts, among all categorical systems, in all possible situations, because the capacity to retain such information is genetically impossible. Computationally, this would lead to "combinatorial explosion." It is impossible for evolutionary inheritance to predict all possible circumstances that a future organism could encounter. The lack of an overarching algorithm to resolve structural conflict among affective hyperpriors is the lack of a meta-prediction.

Moreover, even if such a meta-prediction *were* feasible, it would not do away with uncertainty:

> it is conceivable for an extremely complex set of model algorithms to evolve (no matter how unwieldly they become) which compute relative survival demands in all predictable situations, and to prioritise its action options on

this basis, notwithstanding the 'combinatorial explosion'. But how does the organism choose between *A* and *B* when *uncertainty itself* becomes the primary determinant of action selection? This is what happens in novel situations, for example, which are far from rare in nature. (Solms, 2021b, p. 202, emphasis in the original)

The choice faced by the midbrain decision triangle is not dependent on predictable outcomes. Rather, the driving force is surplus prediction error. The decision is not between different sensory or motor predictions, but between *different categories of uncertainty*. This is why "uncertainty itself" is the primary determinant of action selection. The Lacanian neuropsychoanalytic principle *There is no Homeostasis of Homeostasis* (see Chap. 3) thereby finds consilience with the Solmsian *There is no meta-prediction*—insofar as such a mythical meta-prediction would provide the meta-homeostatic resolution of conflicting homeosta*ses*. This meta-prediction does not exist.

Insofar as predictions operate like signifiers (see Chap. 7), the predictive network can be supposed to operate like a symbolic system. Indeed, Lacan himself proposed that computational and cybernetic communication logic follow symbolic mechanisms (Lacan, 1953–1954). The structural conflict of Pankseppian hyperprior predictions is isomorphic with the Lacanian real as structural antagonism immanent to the symbolic. Innate conflict among affective hyperpriors names a point of the real within the brain.

Antagonism is Necessary for Consciousness

Furthermore, for Solms (2020a), antagonism at the level of affective hyperpriors is not an unfortunate deficit to be removed. It is *necessary* for consciousness. Were a meta-prediction to exist, there would be no uncertainty and therefore no consciousness (recall that affective consciousness is *felt uncertainty*). Rather than a meta-prediction that could resolve all situations of conflict under uncertainty, affective consciousness arises as the "answer" to structural antagonism.

The prioritization function of precision-weight modulation deals with the problem of conflict under uncertainty. However, prioritizing an

8 The Neuronal Real: Antagonism Immanent to the Brain 115

emotional system *does not* provide the perfect dictates to explain the prediction errors at stake. Recall the problematics of motor axis splitting (see Chap. 7) and the general necessity to supplement instincts with experience-dependent learning (see Chap. 6). Prioritizing an emotional system adjusts precision-weights, palpating throughout the predictive hierarchy to allow the organism to *feel its way through uncertainty* by prioritizing a *particular category of uncertainty*. Such precision-adjustments are not guaranteed and are idiosyncratic to individual history: precision itself must be predicted. Simply put, the prioritization function of affective consciousness does not do away with uncertainty or remedy the structural antagonism of conflicting hyperpriors.

Recall Žižek's (2009) formulation that the neuronal real opens onto the psychoanalytic real (see Chap. 3). Insofar as affective consciousness is *jouissance*, here the neuronal real of conflicting affective hyperpriors opens onto the psychoanalytic real of *jouissance*. In the lens of transcendental materialism, Solms's model thereby describes the *emergence* of an irreducible phenomenon from the antagonism of natural inheritance (Johnston, 2019; see Chap. 3).

This Lacanian neuropsychoanalytic lens nuances Solms's characterization of affective consciousness as *feeling one's way through uncertainty*. Crucially, this "feeling"—that is, *jouissance*—is *extimate* to the predictive ego, insofar as affective consciousness is defined as surplus prediction error (see Chaps. 6 and 7). The prioritized emotional system is a particular prioritization of *jouissance* that provides some guidance, but it is a guidance that splits off and follows its own logic. If *jouissance* is the guide for feeling one's way through uncertainty, it is a guide prone to aberration and hooking into the logic of the drive.

This is most evident in the fact that *any* predictions laid down in the wake of the prioritization function will necessarily be *compromise formations* because it is impossible for any prediction to satisfy all conflicting systems, to fully minimize free energy (Solms, 2018). In this sense, *all automatization is premature* (Bazan, personal communication). And yet, these non-declarative predictions will be the organism's precise "solutions." Such a kludge mirrors the structural splitting of the Freudian drive (Johnston, 2005): the impossible demand of drive-source and drive-pressure (here, the impossibility of fully resolving conflicting hyperpriors that gives rise to the pressure of the prioritized emotional system) and the failed attempts of drive-aims and drive-objects to alleviate the tension (here, predictions as compromise formations that fail and, moreover, are charged to repetition via sensitization, precision, and non-declarative mechanisms; see Chap. 7).

Structural antagonism in affective hyperpriors is thereby necessary for a strange consciousness, a consciousness beyond the predictive ego, a force that guides solutions which self-sunder and repeat despite their failures—moving further and further from a meta-homeostasis that never existed to begin with. And yet, we are conscious only because of these warps and antagonisms. Were no such torsional points to exist, there would be no need for consciously feeling one's way through uncertainty. In this Lacanian lens, affective consciousness is quite far from homeostasis as typically conceived.

Affective Consciousness in the Empty Space of Uncertainty: Situating $, *a*, and J

We are now in a position to draw more connections between Lacanian concepts and the present neuropsychoanalytic model. Recall that the Lacanian barred subject ($) is divided by the signifier. Žižek (2020a) explains the retroactive positing of the barred subject:

(1) The subject attempts representation in the symbolic.
(2) This attempt at representation fails.
(3) This failure *is* the subject ($).

This paradoxical logic finds surprising clarity in Solms's model:

(1) Contradicting hyperpriors under conditions of uncertainty demand predictive resolution.
(2) The attempt at predictive resolution fails.
(3) This failure *is* the persistence of affective consciousness as surplus prediction error.

8 The Neuronal Real: Antagonism Immanent to the Brain 117

The Lacanian barred subject indexes an empty space, a space not filled with any particular representation because the subject is precisely what escapes representation. Or, more specifically, the subject is the effect of the impasses of the symbolic. It slips in the cracks between signifiers, never exactly nailed down. I propose that the Lacanian barred subject ($) finds its first[1] roots in the structural antagonism of affective hyperprior predictions. $ names the empty space opened by the lack of a meta-prediction.

Recall that *objet a* is the objectal correlate of the barred subject ($), insofar as the division on the side of the subject mirrors the alluded to excess posited in the representational world (Lacan, 1964; Žižek, 2009). After the midbrain decision triangle prioritizes an emotional system, that system palpates throughout the predictive hierarchy, adjusting precision-weights to guide action and perception. However, these predictive mechanisms still revolve around a core of uncertainty. Prediction errors continue to pass through the hierarchy in a cyclical fashion, a remainder of uncertainty not fully resolved. Some predictions themselves generate spikes of uncertainty (cf. repetition of sensitized non-declarative predictions; see Chap. 7). In other words, the predictive hierarchies under the guidance of a prioritized emotional system are interwoven with uncertainty.

I propose that *objet a* can be used to broadly name the surplus of uncertainty entwined with ongoing predictive work. Recall how the splitting off of the motor axis at the level of premature automatized non-declarative memories is one way to frame the aberrant hooking of instinct into the drive's logic of excess oriented around *objet a* (see Chap. 7). Prematurely automatized memories are *predictions* that *do not work* but nevertheless repeat (Solms, 2018), rendering their own enjoyment (Bazan & Detandt, 2013). Such repetition encircles *objet a within the predictive field* (which includes both perception and action). The persistent uncertainty within predictive cascades corresponds to *objet a* as a *bulge in the predictive field*. Whereas the barred subject ($) drops out as the empty space in the contradiction among hyperpriors, *objet a* indexes the surplus

---

[1] I say "first" because this dialogue with neuroscience demonstrates the need to nuance levels of the symbolic (see Chap. 13).

of uncertainty in the predictive cascade. These are two points on the same moebius strip (Miller, 2023).

These Lacanian concepts allow one to draw out greater nuance within Solms's model. At least two logical moments exist. First, there is the contradiction among hyperpriors in situations of uncertainty. Then a particular system is prioritized. Between these two logical moments is the midbrain decision triangle's prioritization function which is controlled by *predictions of precision*—specifically, which category of uncertainty is inferred to provide the maximal opportunity to minimize uncertainty (see Chap. 6). A logical moment of what I call *uncertainty of uncertainty*—uncertainty over which category of uncertainty to prioritize—precedes the prioritization function and its precision-weight modulation.

I suggest that the distinction between *das Ding* and *objet a*—and between traumatic *jouissance* and titrated *jouissance* bound to signifiers (see Chap. 3)—can help clarify these moments. The over-proximity of uncertainty before which the subject is helpless approximates *das Ding*—not as a concrete substantialization of the real but the real as pure traumatic antagonism—or a pure *jouissance*: J. Selective prioritization of an emotional system transposes this over-proximate *uncertainty of uncertainty* (J) into a *particular category of uncertainty*. Predictive work within the prioritized category of uncertainty gradually (as one moves toward the periphery of the predictive hierarchy; see Chap. 6) shifts the pure traumatic antagonism of the real (here, structurally conflicting hyperpriors) into uncertainty in certain predictive contexts and cascades. At a certain point, this antagonism is worked over into a lack/surplus in the predictive field: *objet a*.

Soler (2015) similarly distinguishes Lacanian "anguish" from others affects like fear, panic, and so on. Anguish—which Soler emphasizes has no predictable behavioral correlate—arises at the over-proximity of the *objet a*, when the *objet* is no longer a veiled, elusive surplus bulging in the phenomenal field. It concerns the overbearing closeness of the *question* of what the Other wants—that is, the overbearing closeness of *uncertainty*. Once there is some minimal sense of what the Other wants, anguish becomes a particular affect.

In the terms presented here, anguish corresponds to the uncertainty of uncertainty, a pure overbearing J indexing the radically unknown dimension of the Other (i.e., *das Ding*; Copjec, 2004; Lacan, 1959–1960; see Chap. 4). Prioritizing a particular category of uncertainty delimits the bounds of the Other's enigmatic presence. It does not explain what the Other wants, but it reduces some of the enigma. In simpler terms, the ungrasped Other (as traumatic J of *das Ding*) becomes an Other to be FEARed, to be PLAYed with, to be enRAGEd with, to bond with (PANIC), to LUST after, to CARE for, and so on. Prioritization corresponds with a particular quality of affect becoming consciously felt and guiding precision-weight modulation to delineate an affective stance toward the predictive field.

Therefore, we can discern logical moments of J and a. To be clear, I am not proposing a mythical pure *jouissance* (J) separate from affective consciousness prioritization (Leader, 2021; see Chap. 11). It is nevertheless helpful to recognize J as a logical moment to draw out a *spectrum* from the contradiction of affective hyperpriors (uncertainty of uncertainty) and the transposition of lack into the predictive field. The prioritization of an emotional system does not immediately transpose J into a in terms of titrating *jouissance* (see Chap. 13). A range can be conceived from J to a as one approaches the periphery of the predictive hierarchy, where greater predictive framing and contextualization titrates and attenuates *jouissance* into a more delineated or localized lack: *objet a*.

The tipping point (and it is a simplification to think of *one* tipping point) likely lies somewhere at the border between non-declarative (automatized) predictions and declarative predictions, due to greater flexibility in declarative predictions, their capacity to generate "mental solids," and the fact that non-declarative predictions *cannot be thought about* in self-reflective consciousness (i.e., they evade the imaginary of the ego). The J of *das Ding* sits closer to the surplus prediction error effected by structural contradiction and premature automatization. *Objet a* points to the remainder or bulge within the predictive field, especially toward the periphery where mental solids dominate.

Again, there is no J outside of affective consciousness—which would risk conceptualizing a reductive meta-affect of sorts. There are only the emotional systems which, when prioritized, operate not in the space of guaranteed homeostasis but in the empty space of subjectivity ($) that follows the logic of excess, of the drive oriented toward a. If one writes

Table 8.1 Lacanian concepts and their proposed neuropsychoanalytic correlates

Lacanian Concept	Neuropsychoanalytic Correlate	Description
Jouissance	Surplus prediction error, affective consciousness	Excess excitation outside predictive/symbolic binding capacities
Barred Subject ($)	Conflicting affective hyperpriors	The empty space in the structural antagonism among affective hyperpriors
Pure J, das Ding	Encounter of structural antagonism of conflicting hyperpriors	Logical moment of uncertainty of uncertainty preceding prioritization function
J(E)	Prioritized emotional system operating in affective consciousness	Jouissance of the prioritized emotional system, enjoyment derived in the system's being hooked into the excess logic of the drive
Objet a	Lack/surplus uncertainty transposed into the predictive field	Uncertainty bulging in the predictive field, especially beyond the hierarchical point of non-declarative predictions; automatized predictions frame objet a

Note: Summary of Lacanian terms situated in neuroscientific terms thus far

one of Panksepp's emotional systems as E_1, then a Lacanian neuropsychoanalytic perspective would write the prioritization of E_1 as $J(E_1)$: *jouissance* of emotional system 1. Including the symbol J highlights that the prioritized Pankseppian emotional system operates in the *hole* of predictable homeostasis and thus follows the excess logic of drive whose repetition derives *jouissance*. There is an enjoyment of the prioritized category of surplus prediction error: $J(E)$.

Because the following chapter will connect more Lacanian concepts to this neuropsychoanlaytic model, Table 8.1 summarizes the links made thus far.

References

Bazan, A., & Detandt, S. (2013). On the physiology of jouissance: Interpreting the mesolimbic dopaminergic reward functions from a psychoanalytic perspective. *Frontiers in Human Neuroscience, 7*, 709. https://doi.org/10.3389/fnhum.2013.00709
Copjec, J. (2004). *Imagine there's no woman: Ethics and sublimation* (2nd ed.). MIT Press.
Dall'Aglio, J. (2021). Sex and prediction error, part 2: Jouissance and the free energy principle in neuropsychoanalysis. *Journal of the American Psychoanalytic Association, 69*(4), 715–741. https://doi.org/10.1177/00030651211042377
Johnston, A. (2005). *Time driven: Metapsychology and the splitting of the drive.* Northwestern University Press.
Johnston, A. (2019). *Prolegomena to any future materialism, Volume two: A weak nature alone.* Northwestern University Press.
Lacan, J. (1953–1954/1991). *The seminar of Jacques Lacan, Book I: Freud's papers on technique* (J.-A. Miller, Ed.; J. Forrester, Trans.). Norton.
Lacan, J. (1959–1960/1992). *The seminar of Jacques Lacan, Book VII: The ethics of psychoanalysis* (J.-A. Miller, Ed., & D. Porter, Trans.) Norton.
Lacan, J. (1964/1978). *The seminar of Jacques Lacan, Book XI: The four fundamental concepts of psychoanalysis* (J.-A. Miller, Ed., A. Sheridan, Trans.). Norton.
Leader, D. (2021). *Jouissance: Sexuality, suffering and satisfaction.* Polity.
Miller, J.-A. (2023). *Analysis laid bare.* Libretto Press.
Soler, C. (2015). *Lacanian affects: The function of affect in Lacan's work* (B. Fink, Trans.). Routledge.
Solms, M. (2018). The neurobiological underpinnings of psychoanalytic theory and therapy. *Frontiers in Behavioral Neuroscience, 12*, 294. https://doi.org/10.3389/fnbeh.2018.00294
Solms, M. (2020a). New project for a scientific psychology: General scheme. *Neuropsychoanalysis, 22*(1–2), 5–35. https://doi.org/10.1080/15294145.2020.1833361
Solms, M. (2020b). Response to the commentaries on the "New Project." *Neuropsychoanalysis, 22*(1–2), 97–107. https://doi.org/10.1080/15294145.2020.1843215
Solms, M. (2021a). Revision to drive theory. *Journal of the American Psychoanalytic Association, 69*(6), 1033–1091. https://doi.org/10.1177/0003065121105
Solms, M. (2021b). *The hidden spring: A journey to the source of consciousness.* Profile Books.

Žižek, S. (2009). *The parallax view*. MIT Press.
Žižek, S. (2020a). *Hegel in a wired brain*. Bloomsbury.
Žižek, S. (2020b). *Sex and the failed absolute*. Bloomsbury.
Zupančič, A. (2017). *What IS sex?* MIT Press.

9

Real, Imaginary, and Symbolic Knottings in the Predictive Model

Abstract This chapter introduces the Lacanian concepts of the Other and the fundamental fantasy, alongside the notion of shared generative models in the Free Energy Principle. This allows me to sketch how the real, imaginary, and symbolic registers are knotted in dynamic predictive processes across the brain's different memory systems. I also develop how the Lacanian symptom can be situated in the brain as deeply automatized motoric predictions soldered to the innate contradictions of affective consciousness.

Keywords Fundamental fantasy • Shared generative models • Other • Lacan • Free energy principle • Predictive coding • Borromean

My first paper on Lacanian neuropsychoanalysis was titled "Of brains and Borromean knots" (Dall'Aglio, 2019). Ironically, that paper did little—if anything—to explain the idea of Borromean knotting or how it might occur in the brain. Now, with the conceptual bridges developed above, it is possible to speak of neuronal Borromean knotting in more

© The Author(s), under exclusive license to Springer Nature Switzerland AG 2024
J. Dall'Aglio, *A Lacanian Neuropsychoanalysis*, The Palgrave Lacan Series,
https://doi.org/10.1007/978-3-031-68831-7_9

detail. This chapter lays out some preliminary points of RSI (real, symbolic, imaginary) knotting in the brain.

The Other, Fundamental Fantasy, and Shared Generative Models

Before I discuss RSI knotting in the brain, it will be helpful to situate the concept of the Other in relation to neuropsychoanalytic predictive coding. The Lacanian capital-O Other (the Big Other) stands for the social order writ large, the repository of signifiers. It is a *position*, not a person—although, individuals can occupy or incarnate this position (Verhaeghe, 2004). For example, when a judge delivers a ruling, they are speaking from the position of the Law-decreeing Other, not just another person or ego. Big Other contrasts with lowercase-o other (the little other), which is the mirror-image to the ego, and other who is *like me*.

Alienation

We are all born into the field of the Other: into family, into culture, into language that precedes us. Lacan distinguishes at least two logical moments in relation to the Other. The first is *alienation*. Helpless before the pressure of the drive, the child dealing with the unbearable real turns to the Other whose actions and speech do something to alleviate drive-tension (see Chap. 4). Certain signifiers take on special weight, the signifiers to which *jouissance* sticks. The subject takes on these signifiers—master signifiers, S1s—and symbolically identifies with them (Lacan, 1964; Laurent, 1995; Zupančič, 2017). Identification with S1s provides symbolic anchoring points, a basis for direction in life. In the case of Mr. A., one can think of "separate" as an S1. This signifier governs Mr. A's way of going through life, his way of dealing with affect and drive.

However, alienation alone is deadening. The subject is completely rivetted to a particular signifier, mortified by the symbol (Lacan, 1959–1960). Laurent (1995) gives the example of "bad boy" as an S1 for a patient. Solely with alienation, this S1 does represent the subject to another signifier—say the mother's discourse which designates the child as such. However, this S1 is alienating insofar as the boy is *just* a "bad boy." He may be a "bad boy," but surely he is more than that. Yet the S1 achieves a pronounced, enigmatic weight, above and beyond other possible signifiers and meanings (S2's). Moreover, with alienation alone, the Other is not lacking; it is a fully determining field of signifiers to which the subject is subjected (Verhaeghe, 2019).

Separation

Lacan describes a second operation: *separation*. Here, S1 is linked to the battery of signifiers (written as S1-S2) and the subject attains an identification not with S1 in the field of the Other but with the *lack* in the field of the Other: *objet a*. The Other is no longer a fully determinate symbolic battery but a figure who themselves is barred, inconsistent, and so on. Rather than the sole repetition of a monotonous S1 as representative of the subject, indeterminacy opens the space for vivifying *jouissance* (Laurent, 1995; Verhaeghe, 2019).

Here is a lack in the signifying chain of S1-S2-S3-…. This is not a lack regarding the meaning of S1; it is a lack inscribed into the Other as the network of signifiers (S1-S2). Separation opens lack on the side of the subject's being and in the field of the Other (Verhaeghe, 2019). This corresponds with the extraction of *objet a* into the Other (Lacan, 1964). Thus, the subject can orient *desire* based on the lack in the Other.

The Fundamental Fantasy[1]

These two operations—alienation and separation—are one way to describe the relationship between the subject and the Other. This relationship takes on an abstract, schematic quality—a generalized template that indicates the subject's stance toward the Other who is not a person but a social position. Lacan (1964) calls this abstract schematic relationship the *fundamental fantasy*.

---

[1] Here I develop arguments from Dall'Aglio (2023a).

Freud's (1919) *A Child is Being Beaten* illustrates the logic of fundamental fantasy. Freud posits three moments in his study of neurotic patients with "beating fantasies":

(1) My father is beating a child and I am looking on.
(2) My father is beating me.
(3) My father (or a father substitute) is beating (or some substitute action) other children while I watch.

Lacan (1956–1957) highlights that the second phase of the fantasy differs from the first and third through its dual structure. The first and third moments include three elements: agent (father or father substitute), object (child or child substitute), and subject (as onlooker). In the second moment, there is only the agent (father) and the subject (as object). The second phase delineates a stance between subject and Other. Notably, the subject is only inserted into the fantasy as an object deriving masochistic enjoyment—which perhaps accounts for the second phase's repression. Lacan highlights that the second phase of the fantasy is cut-off from the other two insofar as it is properly unconscious. The first and third phases are (pre)conscious; the second is unconscious and can only ever be constructed in analysis (Freud, 1919).

The Lacanian fundamental fantasy can be considered the virtual (unconscious, not declarable) terminal point upon which all preconscious fantasies converge (Žižek, 2020). It is an abstract structure that underlies particular manifestations of the fantasy in dreams, transference, symptoms, and so on. It describes the general relationship between the subject and the Other (Miller, 2023; Verhaeghe, 2004).

Lacan (1964) provides the following formula for the fundamental fantasy: $\$ \lozenge a$, the barred subject in relation to *objet a* in the field of the Other. The lozenge ($\lozenge$) indicates the operations of alienation and separation, whereby the subject $\$$ takes on master signifiers from the Other (S1) but also carves an opening toward lack (*a*) in the Other as the battery of signifiers (S1-S2).

I briefly develop the Lacanian Other because the fundamental fantasy is a premier moment of RSI knotting (Miller, 2023). There is the cut-off quality of the unconscious fantasy as a phrase: *My father is beating me*. It is an *isolated axiom* that nevertheless takes on linguistic vicissitudes (i.e., substitution and displacement of the agent, object, and action in the preconscious fantasies). Here is the symbolic dimension of the fantasy as a master signifier: S1. There is also the "freeze-frame" quality of the fantasy as a (constructed) screen-memory and manifesting in preconscious, declarative representations and derivatives (Lacan, 1956–1957). Here is the imaginary element of the fantasy, facilitated by the connection of S1-S2 which makes signification possible. Lastly, there is the element of masochistic enjoyment given shape by the fantasy, the real of *jouissance* incarnated as *beating*. The fundamental fantasy knots these registers together in an idiosyncratic relation to the Other.

Shared Generative Models

This elucidation of the Lacanian fantasy demonstrates how, for Lacan, the subject is densely caught up in the symbolic—inseparable from the relation with the Other regarding key symbolic cornerstones (S1s) and lack as space for *jouissance* (*objet a*). This relationship involves operations (alienation, separation) regarding special signifiers, their network (S1-S2-…), and how this network demarcates *jouissance*. I have already alluded to the idea that predictions can correspond to signifiers. Now let us develop this parallel in greater depth.

Recall that the predictive network is not simply on the side of the subject (see Chap. 7). Predictions derive from the social world, prior experiences, and key interpersonal relationships (Fotopoulou & Tsakiris, 2017; Holmes & Nolte, 2019) and orient relationships to certain types of objects (Kernberg, 2022; Solms, 2017, 2021). Recall Mr. A's fright at the snake. "Separate" was an S1 that came from the Other, incarnated as the man. He used a prediction that was not his own. One could even say that his prioritization of FEAR of the snake (rather than, say, SEEKING) derived from moments in his history—that is, prior circumstances whose traces biased the inference that FEAR-precision-weights offered greater opportunity to minimize free energy than SEEKING-precision-weights. The necessity for precision to be predicted adds to the centrality of the Other in prediction and the prioritization function.

However, one can go a step further. At the level of language and relationships with others, predictions are not mere derivatives from the social world internalized to one brain. Consider the following situation described by Friston:

> In communication and the interpretation of intent, the very notion of theory of mind speaks directly to inference, in the sense that theories make predictions that have to be tested against (sensory) data. Imagine two brains, each mandated to model the (external) states of the world causing sensory input. Now imagine that sensations can only be caused by (the action of) one brain on the other. This means that the first brain has to model the second. However, the second brain is modelling the first, which means the first brain must have a model of the second brain, which includes a model of the first—and so on *ad infinitum*. At first glance, the implicit infinite regress appears to preclude a veridical modelling of another's brain. (Friston & Frith, 2015, p. 130)

Lacan describes a similar problematic of infinite regress as characteristic of the imaginary register (Lacan, 1953–1954, 1954–1955). A child might try to win a game by guessing what the other person is thinking, but the other child might anticipate this and act accordingly—an anticipation which the first child might account for—and so on in an endless back-and-forth between two mirror egos. Such logic exists in an imaginary stance toward the other.

Friston proposes a solution to this problem through the idea of a "shared generative model" and "federated inference." Rather than try to infer the other's predictive model (thus my predictive model includes inferences of the other's predictive model which includes inferences of mine, and so on), the two speakers infer a *common* external state that explains the actions and sensations of *both* speakers. Friston elaborates:

> However, this infinite regress dissolves if each brain models the sensations caused by itself and the other as being generated in the same way. In other words, if there is a shared narrative or dynamic that both brains subscribe to, then they can predict each other exactly—at least for short periods of time. (Friston & Frith, 2015, p. 130)

The hidden state is not placed in the other but rather in a shared space. Each speaker infers that commonly posited hidden state as a common causal explanation. Thus, the generative model formed for the conversation is a *shared* generative model—each speaker's generative model seeks to infer a common causality. When each speaker occupies a limited perspective on this common hidden state, they must somehow communicate their beliefs (likelihood mappings; see Chap. 5) to facilitate maximizing the accuracy of shared predictions in the shared generative model. The necessity for a shared means of expressing beliefs results in the emergence of linguistic systems (Friston et al., 2024). The speakers then collectively minimize free energy through a shared generative model of a common hidden state.

I propose that the Lacanian Other is similar (but not identical) to the idea of a shared generative model. The shared generative model is a position, not a person—a position that nevertheless has its own rules: its structure explains the actions and sensations of subjects who buy into that model. More generally, what would the social world be if not a shared generative model—a *shared* set of predictions that govern the action-perception cycles of *multiple* individuals? Social rules—such as which side of the road to drive on—exist in the space of a shared generative model (an Other) which explains the behaviors of thousands of little others in their little cars. Why do these little others behave like this? There is a law for driving on two-ways roads: a *shared prediction.*

Let us return to Laurent's "bad boy." This patient is not just a "bad boy" to himself; he is also a "bad boy" to his mother. The S1 "bad boy" explains not only his own behaviors (say he makes some transgression) but also the behavior of others (say his mother punishes him). "Bad boy" as a prediction exists not simply *inside* his brain or his mother's, but in a shared space (the Other) of which both brains form a shared generative model.

Fotopoulou and Tsakiris (2017) propose a similar reasoning for a second-person basis of homeostatic selfhood. They argue that the infant's motor prematurity requires that the mother attune to the child's needs—that is, she must *infer* the child's homeostatic state. The child then takes on these predictions. The representation of internal bodily states is a co-creation between mother and infant that explains the sensations experienced by both figures. The Other is this third space, a space of shared inferences, shared predictions, shared signifiers.

Connecting the Lacanian Other to shared generative models highlights how, in a certain sense, the subject's predictive model is Other. Not only is affective consciousness foreign to the predictive ego as the real of *jouissance* within the symbolic; the predictive ego itself is also alienating, being derived from the world of shared predictions. This is why, for Lacan, the subject is divided and empty. There is no true kernel of identity either on the side of the symbolic (which is Other) or the real (structural antagonism). The brain's predictive model is, in a sense, not internal to the brain. It also has an extimate quality, being intimately constitutive of the ego while deriving from an external Otherness. In other words, the brain includes predictions from beyond the Markov blanket of the skin.

The difference between the Lacanian Other—as battery of signifiers—and Fristonian shared generative models and federated inference (Friston et al., 2024) would be what is understood by "communication." Friston's subjects in conversation (see above) are trying to *understand* a common hidden state causing both of their sensory states. Language here is a medium of communication.

Communication is the imaginary dimension of language insofar as it concerns meaning and signification. It is an imaginary relation to the other. Psychoanalysis, on the other hand, has long recognized that language is not principally for communication (Freud, 1895; Lacan, 1953–1954). Recall that the infant's *cry* first effects an *internal change*, only later becoming a means of communication with the caregiving object (see Chap. 4).

The symbolic relation to the Other concerns not meaning but the *materiality* of signifiers. The shared signifiers of the Other are not (first and foremost) shared semantic predictions but shared motoric predictions. Motoric (non-declarative) predictions designate the dimension of the signifier that rivets *jouissance* and represents the subject without production of understanding. In a Lacanian neuropsychoanalytic view, the "material" dimension of the signifier corresponds to the motoric level of predictions: action and thinking as mental action (Bazan, 2011, 2023; Bazan & Detandt, 2013).

Recall the patient who puzzled over the signifier "first" (see Chap. 4). "First" is a shared motoric prediction with a place in the Other. It was a feature of her own predictive model (identified with, as an S1; alienation) but it existed in the shared social space and appeared in specific others as well, such as those saying "Put yourself first." It achieved some particular instantiations (gets linked up to meaning and understanding, S1-S2) but nevertheless remained a puzzling element. The latter is based on the motoric element of the prediction "first" that is insufficiently grasped by semantic predictions. In other words, its network in the Other is discerned as lacking (separation).

Insofar as the fantasy knots the real, symbolic, and imaginary registers and establishes a stance toward the Other, I will use the fantasy to situate RSI knotting within the brain. To do so, I will move from the ground up, so to speak—beginning from the antagonistic bedrock of affective consciousness. While I will repeat some points made in previous chapters, my hope is that this repetition will facilitate some understanding.

Prediction in the Real: Situating Premature Automatization as S1-J

Let us begin from contradictory affective hyperpriors, the logical moment of uncertainty of uncertainty (J). The midbrain decision triangle—based on predictions of precision—prioritizes the system that is expected to best facilitate the reduction of a particular category of uncertainty. Now an emotional system is prioritized in the uncertain space of affective consciousness: J(E).

Despite this conflict among systems, some predictions are laid down and automatized to deal with affective consciousness. Specifically, the prematurely automatized (non-declarative) motoric predictions at the level of basal ganglia are assigned high precision and charged with incentive sensitization, driving a tendency to repeat. These predictions do not eliminate prediction error; they *do not work* and engender their own excitatory repetition.

I propose that prematurely automatized predictions function like an S1. S1 is cut off from the rest of the signifying chain, a signifier all alone whose meaning is unclear. Likewise, prematurely automatized predictions are unconscious, non-declarative—outside semantic grasping. Moreover, they repeat—just as S1 repeats due to its binding of *jouissance*. The excitatory repetition of the prematurely automatized non-declarative predictions procures a bound *jouissance* in repetition. Therefore, at the level of premature automatization, there is real-symbolic knotting, specifically the real of *jouissance* (J) and a master signifier (S1). We can write this as: S1-J.

Moreover, here is a prediction operating (more) in the real. There is a symbolic dimension, but it is not properly symbolic insofar as it is not yet caught up in the system of other signifiers and knowledge (declarative, semantic prediction; S2). S1 is a signifier all alone, a signifier in the real, a signifier that binds something of the real and metabolizes it in some way (Zupančič, 2017).

The Lacanian Symptom

Additionally, I propose that premature automatization—as the knotting of S1-J—corresponds to the Lacanian *symptom*. For Lacan, the symptom is what is most unique and individual to the subject, their singular, contingent way of managing the drive. Indeed, the prematurely automatized prediction forms not in the determinate space of inherited instincts but in the empty space *between* inherited contradiction. Recall the *Fort-Da* example: there is creation in the space *between* PLAY and PANIC (see Chap. 7). In a sense, the symptom is a creation that directly knots signifier and *jouissance* (Miller, 2023; Verhaeghe & Declercq, 2016).

The subject's symptom refers to their fashion of dealing with drive, how they manage *jouissance*. It is a *mode of enjoyment* that *metabolizes jouissance* (Busiol, 2021; Miller, 2023). I contrast *metabolize* with *reduce*. Reducing excitation eliminates that excitation. However, what can one do with the inextinguishable surplus prediction error of affective consciousness—inextinguishable due to the structural antagonism that necessitates a surplus of uncertainty? Here, the subject must "make do" with the tension, find a way to metabolize it, to put it to *use*, and—in the best of cases—derive some enjoyment from it (Miller, 2023). There is always *jouissance* or enjoyment; clinically, one aims to help the patient find a way to enjoy their enjoyment (Fink, 2011; see Chap. 14).

Prediction in the Symbolic: Abstract Structure and the Fantasy[2]

The Lacanian symptom is "nested" within the fundamental fantasy, where the fundamental fantasy provides symbolic mediation or distance from the direct knotting of signifier to *jouissance* (Miller, 2023). Recall the masochistic enjoyment in the discussion of the fundamental fantasy, the real of *jouissance* of *being beaten*. One can take this to be the level of the symptom, a particular motor action (signifier, S1) to which *jouissance* sticks. The unconscious fantasy inserts this mode of enjoyment into a schematic relation with the Other: *My father is beating me*. The shift from symptom to fundamental fantasy involves situating the symptom within an abstract relational structure.

Hierarchical Cognitive Control and Abstract Rules

Automatized action plans do not simply sit idle in the brain. From Bazan et al.'s position, such plans are re-worked into higher-order motor schemas (Bazan, 2011; Jeannerod, 1994). Others (e.g., Badre, 2020) describe how prepotent (habitual, automatized) motor plans are nested within cognitive control hierarchies. This is a developmental process reliant on prefrontal cortex-basal ganglia loops and are largely modulated by dopamine (Ashby et al., 2010; Haber, 2014). Cognitive control refers to the capacity to structure and organize actions to meet goals. Such control mechanisms include inhibition of competing action plans, goal maintenance in the face of distractors (or sub-goals), and selective disinhibition

---

[2] This sections reviews and extends arguments put forth in Dall'Aglio (2023a). I omit in-depth neuroanatomical and functional details in this section, as they are elaborated in the aforementioned article.

of desired action patterns. For example, patients with prefrontal lobe lesions often show deficits in either learning a rule or, once the rule is learned, switching to a new rule despite error feedback—a phenomenon called perseveration (Benson, 1994). This finds a parallel in Freud's (1925) insight that knowledge of the repressed does not undo the process of repression. One may find insight into repetitive habits, but insight alone is often insufficient to *change* the repetition.

Hierarchical cognitive control refers to the idea that control systems are nested within hierarchical rules (Badre et al., 2010; Badre & Nee, 2018). Such rules govern the context-specific implementation of automatized action plans. Consider a child deeply enRAGEd at a classmate out of jealousy for stealing the attention of their friend. Such RAGE might motivate the instinctual action plan to *attack*, but even the sublimated prediction *yelling* is prohibited by the rules of the classroom. A rule has been formed: *if in classroom* (context), *do not yell* (action). An alternative might be formed: *if in classroom, glare*. One will note that the latter rule involves the inhibition of the action plan *yell*. Such inhibition leaves a non-attenuated tension. Such charged traces can be reworked into motor schemas that could still bias specific motor outputs (Bazan, 2011; Jeannerod, 1994).

However, the child is clever and realizes that the teacher is necessary to enforce punishment: *if in classroom and teacher is present, glare*; *if in classroom and teacher is absent, bully and tease*. A rule-context (here, *classroom*) has been nested in a hierarchy (presence of *teacher*). As one progresses up cognitive control hierarchies, rules become increasing abstract, possibly reflecting the internal structure of cognitive control (e.g., arousal-thresholds for competing action plans, policies for implementing rules, etc.; Badre, 2020).

I suggest that the Lacanian fundamental fantasy—as the symbolic mediation of the direct knotting of S1-J—can be understood as an abstract hierarchical cognitive control hierarchy, specifically the hierarchy that governs the context-specific implementation of premature automatized action plans (S1-J). Recall that premature automatized action plans repeat despite the fact that they *do not work*. However, they are not simply implemented in a pure motor repetition (i.e., the *exact* same action plan does not occur repeatedly). They repeat in declarative derivatives in

9 Real, Imaginary, and Symbolic Knottings in the Predictive... 135

a variety of fashions and scenes: in dreams, symptoms, interpersonal relationships, transference, preconscious fantasies, and so on. How are the prematurely automatized non-declarative action plans linked to their varied manifestations in (pre)conscious derivatives?

I propose that abstract hierarchical rules nest the premature automatized motor plan and facilitate its context-specific (i.e., varied) implementation through prefrontal cortex-basal ganglia interactions. Such an abstract hierarchical rule structure also *does not work* in the sense of failing to achieve homeostasis. Just as the premature automatized repressed is a failure (compromise formation) with respect to the impossibility of affective homeostasis (see Chap. 8), the abstract hierarchical control rule (nesting this prematurely repressed prediction and governing its declarative execution) would also be indexed to this affective antagonism. Likewise, it is weighted with consilient precision in a similarly paradoxical fashion, having high confidence to reduce prediction error despite not achieving this. Its high precision would explain its repeated employment as a means of implementing the automatized action plan. Notably, as abstract rules, the predictions involved here would also be non-declarative in the sense of "cognitive habits" rather than motor habits (e.g., how to ride a bike).

The Lacanian fundamental fantasy parallels this abstract hierarchical control rule. Both are unconscious insofar as they are non-declarative. The abstract hierarchical control rule implements the automatized action plan in a specific context, especially interpersonal contexts. In other words, it situates the action plan in the relational world inhabited by the subject by drawing on a fuller network of shared predictions (S1-S2) beyond S1-J. One could say that the fantasy demarcates a *mode of relation* (Busiol, 2021). *My father is beating me* situates the symptom (*beating*) in relation to an Other: *My father*. Through the symbolic mediation of the fundamental fantasy, the *mode of enjoyment* is indexed to an Other who facilitates this enjoyment (Miller, 2023).

Situating Alienation and Separation in Predictive Terms

At this point, the dual operations of alienation and separation can be linked to some neuro-computational terms. There is alienation in the master signifier (S1) that binds *jouissance* in the symptom (S1-J), a *mode of enjoyment* that is rigidified and repetitive (Busiol, 2021). Now with the fantasy, this master signifier (S1) takes on more flexibility in its significations through the rest of the signifying chain (S2). Cognitively, the automatized action plan is nested and linked to other actions and rules, its various instantiations now dependent on semantic, representational, and context-specific input. In other words, the premature automatized prediction (S1) becomes linked to other predictive cascades through the fundamental fantasy as a rule-hierarchy. One can minimally represent this as S1-S2.

Recall, however, that this rule hierarchy, like the premature prediction, *does not work*. This means that the context-specific repetition of the automatized motor plan in the predictive field will still leave some surplus. Excess excitation remains in the instantiations of the motor plan in particular contexts. This excess corresponds to the persistent uncertainty that bulges the predictive field. Here is the persistent non-grasping of *objet a*, the opening in the Other (field of shared predictions) as non-totalized (i.e., separation).

With the fantasy, the connection between motor axis (signifiers) and *jouissance* (*objet a*) is no longer direct. Solms (2018) similarly notes that the premature automatized repressed is not directly accessible to the subject because it cannot be declared in working memory. The connection between signifier and *jouissance* is only graspable through (pre)conscious derivatives, the context-specific implementations indexed to the underlying abstract control rule (i.e., the fundamental fantasy).

As a hierarchical abstract rule governing context-specific implementation of prematurely automatized predictions, the fundamental fantasy achieves a mediated connection between signifier and titrated *jouissance* as *objet a* (Miller, 2023). In other words, the symptom is caught up in the superstructure of the symbolic fantasy-envelope (Verhaeghe & Declercq,

2016). This is a more complex real-symbolic knotting that involves establishing a *mode of relation* (Busiol, 2021) to the Other who facilitates the *mode of enjoyment* (Miller, 2023). Through this complex mediation, there is space regarding the ongoing palpation of affective consciousness as felt uncertainty and the persistent prediction errors in the predictive field. Put in Lacanian algebra: $\$ \lozenge a$. A structural lack on the side of the subject ($\$$; antagonism among conflicting affective hyperpriors) finds relation to lack in the field of the Other (a; the bulge in the predictive field).

Now, we can differentiate between prediction in the symbolic and prediction in the real. Cognitive control orders predictions sequentially through rule hierarchies. They are linked up in a thought-action series. Insofar as signifiers are motoric predictions, one can characterize this series as a *chain* of predictions. At the level of speech, Lacan (1955–1956) characterizes this chain as governed by rules of metonymic combination (grammar, syntax) and metaphoric substitution (word-selection, production of signification). Such rules are controlled by predominantly left hemispheric speech regions (Bazan et al., 2021; Jakobson, 1956; Kaplan-Solms & Solms, 2002). As one moves higher into internal speech and abstract thought (what one might call higher levels of the symbolic; see Chap. 13), these mechanisms rely on increasingly on higher-order prefrontal cognitive control mechanisms (Salas & Yuen, 2016). Higher still, these linguistic rules extend beyond speech and organize behavior and perception in general (Badre, 2020). As an abstract higher order control rule, the fundamental fantasy would include several predictive cascades—that is, several signifying chains that converge onto the same formal structure. These—(motoric) speech and abstract cognitive control—would correspond to prediction operating in the symbolic.

Prediction in the real points to the aspect of the signifier that escapes capture in meaning (imaginary, see below) and hooking onto the signifying chain (symbolic, see above). This would refer to the premature automatized repressed prediction, the S1-J knotting that repeats and insists outside of symbolic mediation. It is a *prediction all alone*.

Prediction in the Imaginary: Image, Understanding, and Declarative Processes

The fantasy also links to the imaginary register, the freeze-frame screen-memory that depicts or constructs the fantasy as a scene (Lacan, 1956–1957). In terms of the model developed above, the imaginary dimension of the fantasy would refer to the context-specific instantiations of the abstract control rule: particular motor, sensory, and affective traces employed in specific experiences (Goetzmann et al., 2018). Such instantiations are imagistic representations of the abstract structure or blueprint. They can be brought to self-reflective consciousness in working memory, presented as "mental solids," and thought through in internal mental space. They are the declarative derivatives of the repressed (Solms, 2018).

The Imaginary in Language: Semantic Meaning

Recall that the imaginary is the register of wholeness, understanding, meaning, and totality (see Chap. 4). In language, the imaginary refers to meaning and signification. More specifically, a knotting of symbolic to imaginary produces signification (Lacan, 1955–1956). Recall that signifiers, for Lacan, lack any determinate connection to a signified. However, everyday experience demonstrates how—at least consciously—referents are typically delimited in speech. This is the function of metaphor, a restriction of (or creation, via substitution) what meaning a signifier might produce (Bazan et al., 2021; Lacan, 1955–1956). Metaphor quilts the signifier to a signified (symbolic to imaginary) to generate signification. Lacan distinguishes this quilting from the syntactic organization of the signifying chain itself, dubbing the contiguous linking of signifier to signifier metonymy (Dall'Aglio, 2023b).

Such linguistic processes are dynamically localizable to the left cortical hemisphere (Bazan et al., 2021; Dall'Aglio, 2019; Kaplan-Solms & Solms, 2002; Salas & Yuen, 2016). Indeed, following Jakobson (1956), Lacan (1955–1956) himself invokes sensory and motor aphasias (neurological syndromes with left perisylvian cortical lesions) to demonstrate the independence of metaphor and metonymy in language (for discussion of Lacan's use of neurology, see Dall'Aglio, 2023b). These regions (including frontal, temporal, and parietal areas) are responsible for both phonological and semantic associative networks. Following Hickok and Poeppel (2004) Bazan et al. (2021) suggest a left hemispheric dorsal-ventral stream model for language processing. Simply put, a ventral temporal path links sound input to conceptual representations through spreading phonological and semantic network activation (Collins & Loftus, 1988). It is an auditory-semantic object-recognition network. A dorsal temporo-parietal path, in contrast, constrains this spreading activation, biasing phoneme selection especially in situations of uncertainty or less "habitual" (i.e., intentional) listening. Recall the example of both *Can certain people win?* and *Cancer in people win*. Phonological and semantic associations are activated upon sensory input (via the ventral stream); a dorsal motoric stream grasps and cuts particular phonological separations and their semantics based on the context. At a peripheral level, (predictions of) the precisions for different predictions guide the cutting of this spreading activation and inhibition of lower-weighted associates. This is an intentional, predictive process: which interpretation best minimizes free energy here?

Prediction in the imaginary refers to the *semantic* domain when concerned with language. Declarative semantic memories involve the meaning attributed to phenomena, the sense made of a sentence for instance. Episodic memories—insofar as language can conjure and inspire egocentric (imagistic) memories of past, present, and anticipated future experiences—would also contribute to the domain of imaginary significations within language. To the extent that a subject generates *declarative meaning* through prediction, prediction is operating in the imaginary.

The Imaginary in the Body-Image

The above discussion focuses on the imaginary in language. However, Lacan first describes the imaginary in terms of the *image*—the specular register of the ego in the mirror (Lacan, 1953–1954, 1954–1955). The little child, helpless before and fragmented by partial drives, sees a totalized image of itself in the mirror—a unity that contrasts with the experience of internal fragmentation. The child identifies (an imaginary identification, an identification with an *image*) with this unity, forming the basis of the ego. Importantly, Lacan's schema does not require a literal mirror. Any number of objects can act like a mirror, insofar as an external image or experience is taken to represent, in a unified way, the child's internal experience (Lacan, 1954–1955). This can involve mirroring as described in attachment theory (Fonagy et al., 2002), whereby the parent interprets and reflects the child's inner state (Verhaeghe, 2004). The imaginary here concerns interpretations of bodily experience—that is, the construction of a body image.

I suggest that the body image refers predominantly to predictions that explain bodily experience in visuo-spatial (including proprioceptive) terms, the specular dimension of the imaginary. Identifying with the image in the mirror demarcates the borders and boundaries of the ego, the bodily limits of "me" and "not-me" (Morin, 2018). Likewise, identifying with the image of the father—*I want to be big and strong like him!*—constitutes an ideal with certain characteristics that have visuo-spatial roots. This would be the dimension where "bad boy" takes on a certain type of body image: clothing, recognizable style of acting, and so on.

I propose that such quilting—between *how* the body-in-space or object acts and *identifying what* this body-in-space or object is—might follow an analogous dorsal-ventral stream system (as described by Bazan et al., 2021) in the right hemisphere.[3] Indeed, the dorsal and ventral stream parallel process model was first proposed for visuo-spatial processing (Milner & Goodale, 1995, 2008). A ventral occipito-temporal object recognition ("what") stream can be dissociated from a dorsal occipito-parietal stream for object location and function ("where"/"how"). Patients with lesions to the ventral stream often present with visual agnosia, a deficit of object recognition, but demonstrate intact knowledge of how to *use* these objects (e.g., the actions associated with turning a key in a lock, without knowledge that this object is a "key"). Likewise, patients with

---

[3] Of course, these systems interact, within and between hemispheres, for visuo-spatial and linguistic processing (Cloutman, 2013). A simple right-left hemispheric distinction for language and visuo-spatial processing is simplistic and ignores the dynamic localization of mental functions (Dall'Aglio, 2019).

lesions to the dorsal stream often present with apraxia, a lack of action-knowledge (or inability to implement action-knowledge) despite intact object recognition (Rizzolatti & Matelli, 2003). The symbolic dimension (dorsal stream, action-knowledge) guides the use and ordering of objects identified imagistically.

Additionally, Lacan's discussions of the imaginary include emotional resonances, especially aggression and eroticism (Lacan, 1954–1955). In a narcissistic fashion, there might be enchantment with an imaginary other in whom one finds imagistic perfection—with the tenuous flip to the inverse aggression that the other threatens my own existence (Soler, 2015). Aggressivity and eroticism can be thought of as Lacan's parallel to Kleinian good and bad part-objects (Busiol, 2021). Specifically, the image takes on a *totalized* affective quality. It is *interpreted* as wholeheartedly positive or negative and *personalized* in an ego, another person. Consider the tyrannical father who prohibits the subject's enjoyment and hoards it for himself. This is the imaginary father cast as a greedy, aggressive other (Lacan, 1956–1957; Israely, 2018).

Symbolic intervention allows the child to nuance and tame the imaginary capture of totalized emotions within the visuo-spatial register: the father is not a tyrannical instantiation of oppression, as he too is subject to the same laws as I (Busiol, 2021; Israely, 2018). Moreover, the intervention of a signifier allows the child to integrate disparate paternal images, where libidinal and aggressive reflections are held together as one person. In somewhat simplified terms, "loving daddy" and "angry daddy" are both "daddy." Altogether, the visuo-spatial imaginary includes the body image as well as a sharp positive-negative emotional split that can be integrated into a Kleinian whole-object relatedness.

Kaplan-Solms and Solms (2002) note that the right cortical hemisphere is especially important for imagistic thing-presentations, the representation of the bodily ego in visuo-spatial terms. They discuss cases of right hemisphere strokes that produce a range of symptoms: left-sided neglect (a bias of attention away from stimuli on the left side of space), left-sided paralysis (due to the contralateral organization of motor control; left-sided motor functions are controlled by primary motor cortex in the right hemisphere), and varieties of anosognosia for hemiplegia. Anosognosia for hemiplegia is a denial of paralysis. Such patients deny that their left arm is paralyzed, responding in a quasi-delusional fashion when confronted with contradictory evidence. For example, if the doctor asked them to move their arm, an anosognosic patient might reply that they *are* moving it. If the doctor notes that the arm is not moving, they might say it is moving in their "mind's eye." If the doctor insists, they might say the arm that is not moving does not belong to them, that is in fact the *doctor's arm*. If the doctor remarks "Then I have three arms?" the patient might say "Yes, that is rather strange!" Varieties of anosognosia include anosodiaphoria, a narcissistic indifference to the paralyzed limb. Such patients will acknowledge their paralysis but might say "I never needed the arm anyway." Another is misoplegia, hatred for the paralyzed limb. The paralyzed limb is recognized but despised as disobedient and faulty (Kaplan-Solms & Solms, 2002).

Kaplan-Solms and Solms (2002) interpret a range of cases of right hemisphere perisylvian lesions in three categories: depression, narcissistic indifference, and misoplegic hatred. They suggest that the common meta-psychological impairment is a capacity for whole-object relations with a concomitant splitting into part-object relatedness. In the case of depression, badness is taken into the ego which refuses to recognize the disappointing (bad, paralyzed) limb. In the cases of narcissistic indifference, intellectualized acknowledgment hides an underlying affective splitting. In the case of misoplegia, the bad-object is manifest and cast out as part of the hated external world. Each example illustrates the entwinement of body-ego boundaries with affective trends (splitting, part-object relatedness).

I suggest that the right hemisphere functions described by Kaplan-Solms and Solms (2002) indicate operations of the imaginary register. Knotting to the symbolic allows a division of "me" and "not me" not purely along lines of positive-negative emotional valence, thereby facilitating whole-object relationships. To be more precise, an intact right hemisphere involves a combination of imaginary and symbolic operations concerning the visuo-spatial register. Given that the left hemisphere language operations also involved imaginary and symbolic dimensions, this is not surprising (cf. dynamic localization; see Chap. 3).

Right hemisphere lesions fray this knotting. With only the mirror-image, the child splits off negative affective experiences and objects into the external world, the operations of the primitive "pleasure ego" (Freud, 1925). Only with the intervention of the symbolic—*that is "you"*—does the child gradually consolidate a "reality ego" that recognizes positive and negative trends *within* the bounds of the ego. The cascade of visuo-spatial predictions designating "you" (both in the perceptual field and regarding what emotional trends belong to you) indexes the imaginary dimension of prediction as linking image to emotion. When the symbolic dimension is damaged, the imaginary persists in a more primitive fashion, splitting at the level of part-objects *and* perception: neglect, anosognosia, and so on.

Morin (2018) adds a Lacanian perspective on right hemisphere patients through a rigorous study of self-portraits. Morin notes that self-portraits of right hemisphere patients with body schema disorders (like those treated by Kaplan-Solms & Solms, 2002) often do not accurately represent the body. Specifically, these types of right hemisphere patients often unilaterally exclude key parts of the body image (e.g., hands, mouth, eyes, clothing).

Morin (2018) proposes that the impairment of the body image coincides with the emergence of *objet a*. Recall that the specular image includes both imaginary mirror-identification and symbolic identification, most rudimentarily seen in the *naming* of the image (an instance of symbolic-imaginary knotting). The specular image coincides with the emergence of *objet a* in the predictive field *as lost*, as the alluded to excess that is not part of the body, the phenomenal bulge. Rather than remaining suppressed, in right hemisphere lesions, this *objet* becomes over-proximate and concretized. Morin (2018) discusses cases where uncanny, quasi-delusional phenomena emerge in right hemisphere body schema patients: the uncanny presence of a unilateral hand, a mouth, an unborn daughter, paranoid suspicions of body imposters, delusions of misidentification (Feinberg, 2010), two limbs on one side, and so on.

How might one think these two perspectives (Kaplan-Solms & Solms, 2002; Morin, 2018) on right hemisphere patients together? It would appear that the emergence of part-object relatedness in a Kleinian lens coincides with the unveiling of *objet a* in a Lacanian lens. I suggest that the asymmetrical presence of certain visuo-spatial objects in Morin's right hemisphere self-portraits (or paranoid and delusional features) are concretizations of *objet a* as a lack in the predictive field. Whereas *objet a* might be sought as the mysterious "it" the other person does or does not have or the enigma in the Other's speech, with the collapse of symbolic capacities (more specifically, *secondary process* symbolic organization; Bazan, 2023), *objet a* shifts from a lack framed *abstractly* to a lack mainly framed imagistically and concretely. The unilateral (missing) hand takes on similar value to the oblong skull in *The Ambassadors* or the silent scream in Edward Munch's (1893) *The Scream*. *Objet a* is concretized in representation as the enigmatic or uncanny surplus.

Regarding part-object relatedness, I suggest that the over-proximity of *objet a* (as *jouissance*) is evident in the intensity of the emotions at play, an arousal beyond the dictates of the "reality principle." Anosodiaphoria indicates a paradoxical emotional stance: indifference toward paralysis, an indifference that—as psychodynamic therapy reveals—narcissistically veils more intense emotions. The poles of depression and misoplegia are (1) excessive emotional trends that are (2) directed in a unilateral fashion toward a split ego and split world. I suggest that these are instances of the surplus in a prioritized emotional system—J(E)—that are closer to the core of the predictive hierarchy (thus less titrated) compared to (symbolically-mediated) whole-object relatedness. Hence the over-proximity of *objet a* and the rudimentary means of framing this negativity.

More could be said regarding the entwining of RSI in language, body image, and object relations. I hope that these above discussions begin to demarcate a general sketch of the terrain while recognizing the complexities involved in RSI knotting in different domains of predictions. Future research will likely shed more light on the particular details unelaborated in the above outline (e.g., the role of the right hemisphere in metaphor as revealed through studies of psychosis; Ribolsi et al., 2015).

Fundamental Fantasy as Knotting Real, Symbolic, and Imaginary

These different levels of the predictive hierarchy reverberate in the fantasy. There are specific, contextual (i.e., declarative, visuo-spatial) instantiations of the fantasy structure. These are imaginary, involving episodic memory, semantic understanding, reflection in working memory, and a visuo-spatial situating of ego and other in the world. Then there is the fundamental fantasy as the non-declarative (unconscious) formal blueprint or rule-structure. This symbolic superstructure nests the real *jouissance* of the symptom.

Now it is clear how the fundamental fantasy accomplishes RSI knotting in the brain. There is (1) the real of the *mode of enjoyment*, S1-J, at the level of premature automatized, incentive sensitized motor traces and action plans, (2) the symbolic framing of this mode of enjoyment as a *mode of relation* to the Other as an abstract stance or way of ordering and organizing experience, and (3) the imaginary instantiations of this abstract schema in particular scenes that can be understood, recognized, imaged, and identified. These moments correspond to moving from the core of the predictive hierarchy (with the $ of antagonistic affective consciousness), toward the symptom (S1-J), toward the fantasy ($◊a), and finally to its declarative, context-specific (imaginary) instantiations. By nesting the symptom and governing context-specific predictions (both through a rule hierarchy), the fundamental fantasy knots the three registers as a mode of relation between $ and the persistent uncertainty (*a*) in the predictive field of the Other.

At this point, several more Lacanian terms have been situated neuropsychoanalytically, so another summary is provided below in Table 9.1.

Table 9.1 Lacanian concepts and their proposed neuropsychoanalytic correlates

Lacanian Concept	Neuropsychoanalytic Correlate	Explanation
Other	Predictive field, especially shared generative models	Other refers to the battery or locus of signifiers, a position that can be occupied. One's signifiers (predictions) come from the Other and occupy a shared space.
Symptom	Premature automatized (incentive sensitized) non-declarative predictions	Sensitized non-declarative action plan that establishes a means of dealing with affective consciousness, a *mode of enjoyment* that knots S1-J.
Fundamental Fantasy	Abstract higher-order control rule that mediates the context-specific implementation of the premature automatized prediction	Unconscious (non-declarative) structural framing of a *mode of relation* to the Other and to *objet a* in the field of the Other (i.e., in the predictive field).

Note: Summary of Lacanian concepts situated in neural terms in this chapter

References

Ashby, F., Turner, B., & Horvitz, J. (2010). Cortical and basal ganglia contributions to habit learning and automaticity. *Trends in Cognitive Sciences, 14*(5), 208–215. https://doi.org/10.1016/j.tics.2010.02.001

Badre, D. (2020). *On task: How our brain gets things done.* Princeton University Press.

Badre, D., Kayser, A., & D'Esposito, M. (2010). Frontal cortex and the discovery of abstract action rules. *Neuron, 66*(2), 315–326.

Badre, D., & Nee, D. (2018). Frontal cortex and the hierarchical control of behavior. *Trends in Cognitive Sciences, 22*(2), 170–188. https://doi.org/10.1016/j.tics.2017.11.005

Bazan, A. (2011). Phantoms in the voice: A neuropsychoanalytic hypothesis on the structure of the unconscious. *Neuropsychoanalysis, 13*(2), 161–176. https://doi.org/10.1080/15294145.2011.10773672

Bazan, A. (2023). Primary and secondary process mentation: Two modes of acting and thinking from Freud to modern neurosciences. *Neuropsychoanalysis.* https://doi.org/10.1080/15294145.2023.2284697

Bazan, A., & Detandt, S. (2013). On the physiology of jouissance: Interpreting the mesolimbic dopaminergic reward functions from a psychoanalytic perspective. *Frontiers in Human Neuroscience, 7*, 709.

Bazan, A., Van de Vijver, G., & Caine, D. (2021). Lacanian neuropsychoanalysis: On the role of language motor dynamics for language processing and for mental constitution. In C. Salas, O. Turnbull, & M. Solms (Eds.), *Clinical studies in neuropsychoanalysis revisited* (pp. 79–104). Routledge.

Benson, D. (1994). *Neurology of thinking*. Oxford University Press.

Busiol, D. (Ed.). (2021). *Lacanian psychoanalysis in practice: Insights from fourteen psychoanalysts*. Routledge.

Cloutman, L. (2013). Interaction between dorsal and ventral processing streams: Where, when and how? *Brain & Language, 127*(2), 251–263. https://doi.org/10.1016/j.bandl.2012.08.003

Collins, A., & Loftus, E. (1988). A spreading-activation theory of semantic processing. In A. Collins & E. Smith (Eds.), *Readings in cognitive science: A perspective from psychology and artificial intelligence* (pp. 126–136). Morgan Kaufmann.

Dall'Aglio, J. (2019). Of brains and Borromean knots: A Lacanian meta-neuropsychology. *Neuropsychoanalysis, 21*(1), 23–38. https://doi.org/10.1080/15294145.2019.1619091

Dall'Aglio, J. (2023a). Extending the theory of premature automatization: The fantasy as an abstract rule in hierarchical cognitive control. *Neuropsychoanalysis, 25*(1), 27–42. https://doi.org/10.1080/15294145.2023.2183888

Dall'Aglio, J. (2023b). Lacan's use of neurology: A neuropsychoanalytic reading of Seminar III, lessons XVII and XVIII. *Lacunae, 25*, 26–73.

Feinberg, T. (2010). Neuropathologies of the self: A general theory. *Neuropsychoanalysis, 12*(2), 133–158.

Fink, B. (2011). *Fundamentals of psychoanalytic technique: A Lacanian approach for practitioners*. Norton.

Fonagy, P., Gergely, G., Jurist, E., & Target, M. (2002). *Affect regulation, mentalization, and the development of the self*. Other Press.

Fotopoulou, A., & Tsakiris, M. (2017). Mentalizing homeostasis: The social origins of interoceptive inference. *Neuropsychoanalysis, 19*(1), 3–28. https://doi.org/10.1080/15294145.2017.1294031

Freud, S. (1895/1966). Project for a scientific psychology. In *The standard edition of the complete psychological works of Sigmund Freud, Vol. 1* (J. Strachey, Ed., Trans.) (pp. 281–391). Hogarth Press.

Freud, S. (1919/1955). 'A child is being beaten.' A contribution to the study of the origin of sexual perversions. In *The standard edition of the complete psychological works of Sigmund Freud, Vol. XVII* (J. Strachey, Ed., Trans.) (pp. 175–204). Hogarth Press.
Freud, S. (1925/1961). Negation. In *The standard edition of the complete psychological works of Sigmund Freud, Vol. XIX* (J. Strachey, Ed., Trans.) (pp. 233–240). Hogarth Press.
Friston, K., & Frith, C. (2015). Active inference, communication and hermeneutics. *Cortex, 68*, 129–143. https://doi.org/10.1016/j.cortex.2015.03.025
Friston, K., Parr, T., Heins, C., Constant, A., Friedman, D., Isomura, T., Fields, C., Verblen, T., Ramsted, M., Clippinger, J., & Frith, C. (2024). Federated inference and belief sharing. *Neuroscience & Biobehavioral Reviews, 156*, 105500. https://doi.org/10.1016/j.neubiorev.2023.105500
Goetzmann, L., Ruettner, B., & Siegel, A. (2018). Fantasy, dream, vision, and hallucination: Approaches from a parallactic neuro-psychoanalytic perspective. *Neuropsychoanalysis, 20*(1), 15–31. https://doi.org/10.1080/15294145.2018.1486730
Haber, S. (2014). The place of dopamine in the cortico-basal ganglia circuit. *Neuroscience, 282*, 248–257. https://doi.org/10.1016/j.neuroscience.2014.10.008
Hickok, G., & Poeppel, D. (2004). Dorsal and ventral streams: A framework for understanding aspects of the functional anatomy of language. *Cognition, 92*(1–2), 67–99. https://doi.org/10.1016/j.cognition.2003.10.011
Holmes, J., & Nolte, T. (2019). "Surprise" and the Bayesian brain: Implications for psychotherapy theory and practice. *Frontiers in Psychology, 10*, 592. https://doi.org/10.3389/fpsyg.2019.00592
Israely, Y. (2018). *Lacanian treatment: Psychoanalysis for clinicians.* Routledge.
Jakobson, R. (1956/1990). Two aspects of language and two types of aphasic disturbances. In L.R. Waugh & M. Monville-Burston (Eds.), *Roman Jakobson: On language* (pp. 115–133). Harvard University Press.
Jeannerod, M. (1994). The representing brain: Neural correlates of motor intention and imagery. *Behavioral and Brain Sciences, 17*, 187–245. https://doi.org/10.1017/S0140525X00034026
Kaplan-Solms, K., & Solms, M. (2002). *Clinical studies in neuro-psychoanalysis: Introduction to a depth neuropsychology* (2nd ed.). Karnac Books.
Kernberg, O. (2022). Some implications of new developments in neurobiology for psychoanalytic object relations theory. *Neuropsychoanalysis, 24*(1), 3–12. https://doi.org/10.1080/15294145.2021.1995609

Lacan, J. (1953–1954/1991). *The seminar of Jacques Lacan, Book I: Freud's papers on technique* (J.-A. Miller, Ed.; J. Forrester, Trans.). Norton.
Lacan, J. (1954–1955/1991). *The seminar of Jacques Lacan, Book II: The ego in Freud's theory and in the technique of psychoanalysis* (J.-A. Miller, Ed.; S. Tomaselli, Trans.). Norton.
Lacan, J. (1955–1956/1997). *The seminar of Jacques Lacan, Book III: The psychoses* (J.-A. Miller, Ed., R. Grigg, Trans.). Norton.
Lacan, J. (1956–1957/2021). *The seminar of Jacques Lacan, Book IV: The object relation* (J.-A. Miller, ed., A.R. Price, trans.). Polity.
Lacan, J. (1959–1960/1992). *The seminar of Jacques Lacan, Book VII: The ethics of psychoanalysis* (J.-A. Miller, Ed., & D. Porter, Trans.) Norton.
Lacan, J. (1964/1978). *The seminar of Jacques Lacan, Book XI: The four fundamental concepts of psychoanalysis* (J.-A. Miller, Ed., A. Sheridan, Trans.). Norton.
Laurent, É. (1995). Alienation and separation (I). In R. Feldstein, B. Fink, & M. Jaanus (Eds.), *In Reading seminar XI: Lacan's four fundamental concepts of psychoanalysis* (pp. 19–28). State University of New York Press.
Miller, J.-A. (2023). *Analysis laid bare*. Libretto Press.
Milner, A., & Goodale, M. (1995). *The visual brain in action*. Oxford University Press.
Milner, A., & Goodale, M. (2008). Two visula systems re-viewed. *Neuropsychologia*, *46*(3), 774–785. https://doi.org/10.1016/j.neuropsychologia.2007.10.005
Morin, C. (2018). *Stroke, body image, and self-representation: Psychoanalytic and neurological perspectives* (K. Valendinova & C. Morin, Trans.). New York: Routledge.
Munch, E. (1893). *The scream (oil, tempera, pastel, and crayon on cardboard)*. National Gallery and Munch Museum.
Ribolsi, M., Feyaerts, J., & Vanheule, S. (2015). Metaphor in psychosis: On the possible convergence of Lacanian theory and neuro-scientific research. *Frontiers in Psychology*, *6*, 664. https://doi.org/10.3389/fpsyg.2015.00664
Rizzolatti, G., & Matelli, M. (2003). Two different streams form the dorsal visual system: Anatomy and functions. *Experimental Brain Research*, *153*(2), 146–157. https://doi.org/10.1007/s00221-003-1588-0
Salas, C., & Yuen, K. (2016). Revisiting the left convexity hypothesis: Changes in the mental apparatus after left dorso-medial prefrontal damage. *Neuropsychoanalysis*, *18*(2), 85–100. https://doi.org/10.1080/15294145.2016.1219937

Soler, C. (2015). *Lacanian affects: The function of affect in Lacan's work* (B. Fink, Trans.). Routledge.

Solms, M. (2017). Some innate predictions are social in nature: Commentary on "Mentalizing homeostasis" by Fotopoulou and Tsakiris. *Neuropsychoanalysis, 19*(1), 55–57. https://doi.org/10.1080/15294145.2017.1309622

Solms, M. (2018). The neurobiological underpinnings of psychoanalytic theory and therapy. *Frontiers in Behavioral Neuroscience, 12*, 294. https://doi.org/10.3389/fnbeh.2018.00294

Solms, M. (2021). *Response to Otto Kernberg*. *Neuropsychoanalysis, 23*(2), 115–119. https://doi.org/10.1080/15294145.2021.1984284

Verhaeghe, P. (2004). *On being normal and other disorders: A manual for clinical psychodiagnostics* (S. Jottkandt, Trans.). Other Press.

Verhaeghe, P. (2019). Lacan's answer to alienation: Separation. *Crisis & Critique, 6*(1), 364–388.

Verhaeghe, P., & Declercq, F. (2016). Lacan's analytic goal: Le sinthome or the feminine way. *Psychoanalytische Perspectieven, 34*(4), 1–21.

Žižek, S. (2020). *Hegel in a wired brain*. Bloomsbury.

Zupančič, A. (2017). *What IS sex?* MIT Press.

Part III

Developing Implications of a
Lacanian Neuropsychoanalysis

10

The Critique of *Jouissance*

Abstract While *jouissance* is often invoked as a foundational Lacanian idea, it has been criticized on both clinical and theoretical grounds as overly simplistic. Darian Leader has charged the Lacanian use of *jouissance* as theoretically imprecise, obscuring the relation to the Other, and smuggling in problematic substantial and energetic presuppositions. To the degree that my Lacanian neuropsychoanalysis relies on *jouissance*, these criticisms pose a potential problem to my meta-neuropsychology. Here, I develop Leader's criticisms of *jouissance* in relation to my Lacanian neuropsychoanalysis.

Keywords Darian Leader • Energetics • Relational • Reductionism • Lacan • Affect

Although *jouissance* is considered a quintessential Lacanian concept, one of the defining features of Lacanian psychoanalysis and a cornerstone of theoretical complexity, it is not without criticism. Darian Leader (2021)

Here I extend arguments from Dall'Aglio (2023).

extensively critiques various deployments of *jouissance* by Lacan and Lacanians more broadly. Given that my Lacanian neuropsychoanalytic project revolves heavily around *jouissance*, his criticisms could be extended to the model developed here.

Here, I review Leader's critique of *jouissance*. I adapt his arguments against *jouissance* in Lacanian psychoanalysis to the specifics of the present Lacanian neuropsychoanalytic model. In the following chapter, I will respond to these criticisms. This allows me to clarify features of my Lacanian neuropsychoanalysis and nuance theoretical points. Moreover, it serves as an illustrative example of how dialogue with neuroscience can contribute to issues within Lacanian psychoanalysis.

Theoretical Imprecision

Leader (2021) claims that *jouissance* is a theoretically imprecise term, often used in a blanket fashion for all bodily and affective phenomena. Lacanian aphorisms such as *jouissance* as "the only substance" restrict how clinicians might interpret phenomena with different qualities and vicissitudes (Leader, 2021, p. 6). He argues "that we are better served by a plurality of concepts rather than one catch-all term, which risks obscuring and covering over important differences in matters both clinical and conceptual" (Leader, 2021, p. 7). One is reminded of the saying: if you only have a hammer, all you see are nails (to which a friend told me the punchline: "and then you're screwed!"). If all you have is *jouissance*, then all bodily innervations and emotional experiences are simply matters of excessive enjoyment.

A similar sentiment appears in the Lacanian privileging of anxiety. Žižek's quip is well-known: "The only emotion which doesn't deceive is anxiety. All other emotions are fake" (Žižek, in Fiennes, 2006). Anxiety, as the common conscious experience of *jouissance*, receives special attention because it supposedly best indexes the real. Other emotions are disguised by imaginary-symbolic colorings, potentially leading the clinician astray into an egoic (defensive) discourse. For some, *jouissance*-as-anxiety is the Lacanian royal road to the unconscious truth of the subject (Fink, 2011; Soler, 2015). Such a theoretical and clinical approach, however, can lead to an ossifying view of *jouissance*. For Leader, theorizing a "One" of *jouissance* "risks substantialising the very ideas and feelings that in the practice of psychoanalysis we aim in fact to desubstantialise" (Leader, 2021, p. 133).

Leader charges Lacanians with referring to *jouissance* as a catch-all for all affective phenomena, effectively overshadowing important nuances. For example, the famous Freudian "strange satisfaction" of the symptom is often invoked as a Freudian antecedent of Lacanian *jouissance*. However, Leader notes that this symptomatic satisfaction is, for Freud,

> a result of a rather complicated process, starting from frustration, moving through phantasy and ending in the construction of symptoms, construed as a form of sexual activity that includes the very forces that work against such activity…[Freud's students argue that] frustration will generate hatred and rage that will mix with or create libido. (Leader, 2021, pp. 12–13)

Such reasoning epitomizes Leader's position. He wants to retain "libido itself as a hybrid of attachment and repulsion, absorption and destruction, preservation and cancellation" to highlight the "fundamental dimension of hybridity" in libido (Leader, 2021, pp. 14, 15). Rather than replace these dynamics with the single signifier "*jouissance*," Leader hopes to preserve a "plurality of concepts" in the service of theoretical and clinical rigor.

This criticism might be adapted to the present Lacanian neuropsychoanalytic model by suggesting that I have replaced the "plurality" of Solmsian-Pankseppian drives with *jouissance*. In highlighting the contradictions of affective hyperpriors and the emergent surplus prediction error (J), I could be charged with ignoring the unique dynamics of each emotional system. Furthermore, I simplify premature automatization as S1-J. This implies a single automatized prediction rivetted to surplus prediction error, without consideration of the plurality of either multiple automatized action plans or different affective trends involved (Lackinger, 2020). Moreover, linking *jouissance* to surplus prediction error could risk obscuring differences among prediction errors occurring at different levels of the predictive hierarchy. *Jouissance* risks simplifying Pankseppian emotions while also becoming too straightforwardly equated with any conscious emotion:

> Where jouissance had once indexed a disturbing beyond to pleasure…today it is often synonymous with, precisely, any private, or indeed shared, pleasure." (Leader, 2021, p. 72)

Jouissance Overshadows the Relation to the Other

Leader further criticizes the idea of auto-erotic, masturbatory *jouissance*—the "pure One of *jouissance*" logically prior to the symbolic—as obscuring the connection between subject and Other (Leader, 2021, p. 9). A Lacanian analyst might interpret bodily tension in an isolated fashion of self-contained enjoyment. For example, consider a patient covered in tattoos. A Lacanian interpretation could easily sound something like: "Tattoos are a way of directly intervening on bodily jouissance, targeting the real of the body without recourse to the Other." Instead, Leader argues that such phenomena might instead represent attempts at separation from or relating to the Other. Tattoos may be an auto-erotic attempt to deal with the drive; or tattoos might be symbolic reminders of previous treasured experiences and people. Seemingly self-contained bodily pleasures may still be relational. Reviewing the writings of post-Freudian analysts, Leader notes that

> states of bodily—and especially genital—tension are not simply experienced as isolated quanta of excitation but as indexes of the Other's presence and absence. If the Other has the power to quell certain bodily sensations—say, by feeding, soothing, etc.—it must have the power to quell others and, in particular, those that cannot be readily treated through manual manipulation or movement. The persistence of the sensation, then, includes within it the reproach to the Other, to the point of becoming indistinguishable. As an analysand explained, the feeling of burning engorgement in her sex was "identical" with her fury at her absent boyfriend for having caused this. (Leader, 2021, p. 17)

Privileging *jouissance* can mask how a latent relational matrix rests behind seemingly isolated body phenomena (valorized as *jouissance* of the body or of the real). Affective excitation is deeply relational, existing in a dialectical interplay between subject and Other. *Jouissance* might be used by some authors to refer to this blending of feelings and interpersonal bonds. However, the term still "fails to account for their genesis" (Leader, 2021, p. 23).

This argument can be adapted to the predictive focus of the present project. Even with the invocation of shared generative models and the predictive field as the Other (see Chap. 8), my model could be characterized as privileging the *jouissance* (as surplus prediction error) that drops out of prediction—that is, not caught up in the predictive field of the Other. This might especially be the case insofar as I carve a logical moment (J) prior to prioritization [J(E)] that only gradually shifts into *objet a* in the predictive field toward the periphery of the hierarchy.

The Jelly-Like Substance and Energetics

Furthermore, *jouissance* as excess necessarily relies on some notion of a "limit" or "boundary" to define an excessive beyond. While Lacanians are critical of homeostasis, Leader points out that they are surprisingly reliant on it when conceptualizing *jouissance*:

> Lacanians do not really have a theory of pleasure, and when asked to define it will invariably just cite Freud's idea of a minimum level of tension…Jouissance then becomes anything 'too much', any excess that threatens the reign of homeostatic balance. (Leader, 2021, p. 74)

Leader reviews discussions by Freud and post-Freudians that complicate this understanding of pleasure as tension-reduction:

> The psyche strived for discharge of excitation, with pleasure equated with a minimal, or at least constant, level of tension. Yet…neither pleasure nor pain could be identified so neatly with the vicissitudes of discharge.

Reductions and increases in tension were also surely not equivalent to the affective states of pleasure and pain. (Leader, 2021, p. 28)

Indeed, my own deployment of *jouissance* relies almost entirely on the idea of a "too much" outside of homeostatic (or predictive) capacities.

Beyond criticizing the view that tension-reduction equals pleasure and tension-increase equals unpleasure (a model which Solms retains), Leader notes that the metaphor of a quantity that can be increased or decreased invokes *jouissance* as a jelly-like substance. One can imagine *jouissance* squeezed from one point to another in displacement, squished in condensation, or perhaps sculpted into a new shape in substitution. It sounds like an unfortunate Jello-cake splashing about like the wandering uterus of ancient Greek hysteria.

Common Lacanian metaphors—*jouissance* captured in the symptom, localized in the Other, drained through speech—paint a misleading picture of *jouissance* as some goo internal to the psyche or body that can be moved about. For Leader, this issue is entwined with implicit Freudian energetic metaphors. Instead, Leader claims:

> It would surely be more correct to say that the patterns, rhythms and processes that generate certain forms of innervation and activation are affected by language and relational structures: 'jouissance' is thus less a thing hidden inside of us...than a product" (Leader, 2021, p. 105)

This criticism could be adapted to this current project by taking issue with the free energy principle. In a certain reading (although incorrect, see Chap. 11), the Free Energy Principle (re)introduces energetics and substantializing metaphors. Free energy or prediction error can be increased, decreased, and so on. Likewise, I rely on the metaphor of a "shift" from *jouissance* in the register of *das Ding* (traumatic over-proximity of uncertainty) to *objet a* as *jouissance* in the predictive field of the Other. This implies a fixed amount of *jouissance* distributed across a spectrum from *das Ding* to *objet a*. Neuropsychoanalysis remains reliant on the notion of quantity.

References

Dall'Aglio, J. (2023). Jouissance and affective neuroscience: A critical neuropsychoanalytic integration. *Scandinavian Psychoanalytic Review*. https://doi.org/10.1080/01062301.2023.2284514

Fiennes, S. (Director). (2006). *The Pervert's Guide to Cinema* [Motion Picture].

Fink, B. (2011). *Fundamentals of psychoanalytic technique: A Lacanian approach for practitioners*. Norton.

Lackinger, F. (2020). Commentary on Mark Solms' "New Project for a Scientific Psychology". *Neuropsychoanalysis, 22*(1–2), 77–80. https://doi.org/10.1080/15294145.2021.1878614

Leader, D. (2021). *Jouissance: Sexuality, suffering and satisfaction*. Polity.

Soler, C. (2015). *Lacanian affects: The function of affect in Lacan's work* (B. Fink, Trans.). Routledge.

11

A Neuropsychoanalytic Contribution to Debates over *Jouissance*

Abstract Here I demonstrate how my Lacanian neuropsychoanalytic formulation of *jouissance* overcomes Darian Leader's criticisms of the concept. Specifically, the brain's predictive logic is unthinkable without the social domain of shared generative models (the Other). Additionally, the Free Energy Principle allows neuropsychoanalysis to furnish an informatic (non-substantial) model of qualitatively distinct domains of (affective) uncertainty. Yet, these lenses of neuroscience are difficult, if not impossible, to conceptualize without retaining Lacanian concepts (including *jouissance*). I thereby argue for the ongoing (but precise) inclusion of *jouissance* in Lacanian (neuro)psychoanalysis.

Keywords Information • Free energy principle • Neuroscience • Lacan • Shared generative models • Affect • Energetics • Relational

Lacanian Concepts Should not Replace Neuropsychoanalytic Ones

Regarding Leader's criticism of using *jouissance* as a catch-all term, I emphasize that *jouissance* should not be used to replace any ideas in neuropsychoanalysis. *Jouissance* should not be used instead of affective consciousness, SEEKING, prediction error, and so on. Likewise, my concept of drive as instinct aberrated into the logic of excess should not be simplified into *objet a*. Hence, I have retained both sets of concepts in the model developed in Section II.

Jouissance, *objet a*, $, and so on are concepts. These ideas find correlates when mapped in neural space which then reconfigures how we think about ideas in neuropsychoanalysis (see Chap. 3). In order to retain this reconfiguration, these neuroscientific ideas must remain in a dialectical interplay with Lacanian concepts. Likewise, these Lacanian concepts have been updated through their dialogue with neuroscience. Relying only on one set of concepts would obscure the theoretical developments made here.

For example, I have suggested that the roots of $ be situated in the empty, contradictory space of affective hyperpriors. This is an *addition* of a concept that only retains its Lacanian neuropsychoanalytic specificity by retaining all the Pankseppian hyperpriors. Likewise for *jouissance*. What I designate as the "pure J" of *das Ding* is a logical moment. In line with Leader, I am not positing some monistic J-elly divided among seven emotional systems. There are *only* the emotional systems and the contradictions among them. The idea of pure J allows one to isolate a key point in Solms's model: the point of uncertainty of uncertainty, before a particular category of uncertainty is prioritized. It is important to isolate this moment because it allows one to ask: what structures the predictions of precision? (I develop this answer in Chap. 12.)

With the Lacanian neuropsychoanalytic matheme J(E), I indicate how the prioritized emotional system does not follow a simple homeostatic logic because it is operating in the empty space of $ (affective contradiction) and follows the excess-logic of drive oriented toward *objet a*. It is a reminder that, regardless of the homeostatic leanings of the emotional system, there is a tendency toward obscene enjoyment *within* that system that leads it astray from uncertainty-reduction ideals. One could rewrite this term depending on the system in question: J(SEEKING), J(RAGE), J(LUST), J(PLAY), and so on.

Importantly, I am not simply dividing seven different types of *jouissance*. Rather, J(E) emphasizes the perverse tendency toward repetition and excess that aberrates each instinct off its homeostatic course. Particular qualities of these aberrations must be investigated for each system within each unique individual. There is no reason to believe that the dynamics of aberration would be identical for each system's different circuitries, neuromodulator volleys, and predictive histories. Indeed, these systems have distinct operations that impact different levels of predictive hierarchies in different fashions (Henderson, 2023; Hoşgören-Alıcı et al., 2023). Likewise, there are likely neurobiological differences in sensitization mechanisms for different neuromodulatory systems. SEEKING, RAGE, CARE and so on are not merely semantic descriptors; they designate specific neurobiological circuits, neurochemistries, and the like.

Retaining the idea of *objet a* as the "bulge" in the predictive field highlights how, even at the level of declarative representation, there remain certain excesses or veils that allude to some beyond. Prediction error is embedded in the phenomenal experience of the perceptual world. This does not mean that different varieties of predictions (e.g., motor, linguistic, semantic, visuo-spatial) can be simply reduced to "prediction" and their attendant surpluses reduced to "*objet a*." Indeed, Lacan (1964) himself differentiated appearances of *objet a* (gaze, voice, breast, and feces) concerning different drive registers (scopic, invocatory, oral, anal). Lacanian terms *add* to (not replace) neuropsychoanalytic ones, and vice versa—while concepts on both sides benefit from the opportunity for substantive revision.

In my view, the discussion of prediction across different registers (especially symbolic-imaginary mechanisms in linguistic versus visuo-spatial domains; Chap. 9) does not only illustrate the applicability and benefit of Lacanian concepts to neuropsychological phenomena. It necessitates diversification of these Lacanian concepts. This does not return us to crude localizationism, but a dynamic knotting of different dimensions of the registers.

The Centrality of the Relation with the Other

Although Leader's "relational" emphasis is not identical with relational psychoanalysis, it is important to emphasize that current advances in neuropsychoanalysis include relational additions (Dauphin, 2023; Salas et al., 2021). Moreover, Panksepp's emotional systems cast the world with certain types of objects (Kernberg, 2022; Solms, 2021). RAGE prioritization renders an object frustrating, PANIC prioritization renders an object caregiving, FEAR prioritization renders an object threatening, and so on. In this way, Pankseppain hyperpriors already index a certain Other—albeit an Other that is non-declarative.

This is why Solms states: "Some innate predictions are social in nature" (Solms, 2017, p. 55). One cannot remove the Other from J as surplus prediction error. What I characterize as the fundamental fantasy creating a *mode of relation* to the Other is the formation of a *patterned* relation to the Other. To say that, at the innate level of conflicting hyperpriors, there is no innate fantasy does not mean there is no space of the Other.

Notably, highlighting "relationality" as a relation to the Other adds a Lacanian twist to relational perspectives in neuropsychoanalysis. Relationality is not merely toward important others. The subject relates to the Other as a general field, the symbolic space that has its material weight in the shared predictions into which egos and others are inserted. This position is occupied by but not reducible to particular others.

Consider the following. Fotopoulou and Tsakiris (2017) rely on three principles for their second-person model of caregiver-infant homeostatic mentalization:

> (1) the progressive integration and organization of sensory and motor signals constitutes the foundations of the minimal self, a process which we have linked to contemporary, computational models of brain function and named "embodied mentalization"; (2) interactions with other people are motivated and constrained by the same principles that govern the "mentalization" of sensorimotor signals in the individual—and hence the mentalization of one's body can include signals from other bodies in physical proximity and interaction, especially in interaction with particular bodies. (3) Crucially, given the dependency of humans in early infancy, there is a

"homeostatically necessary" plethora of such embodied "proximal" interactions, especially as regards interoception. (Fotopoulou & Tsakiris, 2017, p. 3)

Embodied mentalization refers to the idea that the "most minimal aspects of selfhood, namely the feeling of being an embodied, agentive subject, are fundamentally shaped by embodied interactions with other people in early infancy and beyond" (Fotopoulou & Tsakiris, 2017, p. 6). The infant's capacity to regulate hunger, for example, *depends* on sensorimotor interactions with the primary caregiver. Fotopoulou and Tsakiris propose that these embodied interactions form the basis of minimal selfhood, insofar as interoceptive predictions about the self are formed in dyadic pairing with the caregiver.

In computational terms, this is an instance of shared generative models, generalized synchrony, and federated inference (see Chap. 9; Friston & Frith, 2015; Friston et al., 2024). Commenting on Fotopoulou and Tsakiris's model, Friston elaborates:

Indeed, if we can only communicate or interact with things like us, then generalized synchrony—in the setting of active inference—provides an obvious solution to the problem of dyadic communication and neuronal hermeneutics…When cast in the interoceptive domain, I would imagine exactly the same arguments apply. In other words, there will be an inevitable mimicry and synchrony in joint action that enables the acquisition of generative models…The insight afforded by the notion of generalized synchrony is that the infant's generative model comprises narratives that are shared in any (affiliative or interoceptive) interaction with another. Put simply, from the point of view of the generative model, self and [m]other are singing from the same hymn sheet. The only outstanding thing that needs to be inferred is: who is the current protagonist or agent in a turn-taking sense…me or you? Clearly, to make this inference it is necessary to have a mentalization of the difference between self (me) and other (you) that can suitably contextualize the shared narrative. (Friston, 2017, p. 45)

In our terms, differentiating "me" and "you" occurs through symbolic prediction, which allows us to read Friston's commentary in unique light. There are not two elements but rather three: self, (m)other, *and* the "shared hymn sheet," the shared generative model. These correspond to ego, alter-ego, and Other. This difference is not clear in the computational model because it applies the same model for predictions at different hierarchical levels (interoception, communication, etc.). However, when one places special weight on motoric signifier-predictions (Bazan, 2023), the non-semantic level of motoric predictions indicates how, in addition to a communication axis of understanding between self/ego and mother/alter-ego, there is the motoric unconscious axis between subject ($) and Other. In Lacanian terms, this is the L-schema: the communication axis between self and (m)other being the imaginary and the motoric unconscious axis between $ and Other being the symbolic (Lacan, 1954–1955). This highlights the empty space of $ (as opposed to the predictive ego, the inferred minimal selfhood in Fotopoulou and Tsakiris's argument) in relation to the Other *as the predictive field* itself, in its materiality of shared predictions.

Such materiality is exemplified in the slip of the tongue, a motoric prediction that is executed in contrast to the self-reflectively conscious intentions. The slip, surprising, indexes the subject ($) insofar as psychoanalysis does not treat the slip as a mere accident but indicative of something more. For the slip to have this potential, there must have been other—unconscious, non-declarative—rules and motor schemas (Bazan, 2011) that biased its execution. Such biasing reflects the influence of the fundamental fantasy built upon antagonistic hyperpriors (J).

This example returns us to Leader's critique. The Lacanian neuropsychoanalytic emphasis on the shared generative model as Other indicates how moments of surprise (e.g., the slip) are points of inconsistency and excess in the Other. These are coordinates of demarcating *objet a* in the field of the Other. To speak of *jouissance* as surplus prediction error (i.e., as surprise) at this level thus does not obscure the Other. Computationally speaking, there is no prediction error without prediction; there is no *jouissance* without the Other. Even at the level of innate Pankseppian hyperpriors, a minimal social orientation is sustained (Solms, 2017).

Neuropsychoanalysis Replaces Energetics with Information

Fortunately, Leader already begins to supply the answer to his critique of *jouissance* as a jelly-like energetics:

> A further widely discussed reformulation of the energy model was taken from cybernetics. Freud's famous identity of perception could be seen from the perspective of an exchange of energy, but wouldn't it be more accurate, it was argued, to see it in terms of an exchange of information? It was less the volume of excitation circulation in a closed system that mattered here than the process of information encoding and transmission…The idea was basically that the effect of stimulation was less to increase the rate of discharge—so the old tension model—than to impose an order and patterning on it; that is, to encode it. (Leader, 2021, pp. 31–32)

Leader points to a cybernetic alternative to an energetic conception of libido, one that privileges "information encoding and transmission" over jelly-like displacements and discharge. He continues: "if one thing *means* another, it may be experienced *as* the other, and this is a signifying process rather than a quantitative energetic one" (Leader, 2021, p. 106, emphasis in the original).

This is the precise shift introduced by the notion of variational free energy in place of thermodynamic entropy (Friston, 2010; see Chap. 5). Solms (2020a) makes this move in rewriting Freud's *Project* through the Free Energy Principle and *informatic* surprise, free energy minimization as a predictive (inferential, interpretive) process, and so on. Solms clearly states:

> In physics, matter is no longer a fundamental concept; it is an energy state (hence $E=MC^2$). Variational free energy is a function of information exchange (not thermodynamic exchange) between a system and its environment…"Information" is a physical concept. (Solms, 2020b, p. 99)

Note that the "energy state" that underlies matter is not a substantial energetics (i.e., "not thermodynamic exchange") but free energy, an "information exchange."

Nevertheless, Leader's critique holds merit insofar as the tendency to substantialize the brain still prevails in contemporary science. A Lacanian neuropsychoanalysis is perhaps helpful to counter this tendency through a focus on the symbolic and the brain's predictive processes as symbolic and indexed to shared generative models. The notion of shared generative models stresses how the brain's predictive model extends beyond the physical boundaries of the brain's tissue. It also lends weight to a particular interpretation of the Lacanian registers whereby the real is not some substantial Thing external to the symbolic but the name for the point of antagonism immanent to the symbolic order (Žižek, 2020; Zupančič, 2017). This is why, for Lacan, sex is a negativity, the immanence of antagonism around which the subject situates itself (Copjec, 2015).

I hope that it is clear how this Lacanian neuropsychoanalytic model updates and gives precision to Lacanian concepts in such a way that improves theoretical rigor and avoids critiques of *jouissance* made by Leader. However, in responding to these critiques, one may notice a crucial theoretical point. Against the criticism that J obscures the Other, I argued that the Other is already implied at the level of affective consciousness. However, recall that the Other is the battery of signifiers, and signifiers are what tame and bind *jouissance*. In other words, I am claiming that there are signifiers at the affective level of *jouissance* as surplus prediction error. How do I reconcile this?

References

Bazan, A. (2011). Phantoms in the voice: A neuropsychoanalytic hypothesis on the structure of the unconscious. *Neuropsychoanalysis, 13*(2), 161–176. https://doi.org/10.1080/15294145.2011.10773672

Bazan, A. (2023). Primary and secondary process mentation: Two modes of acting and thinking from Freud to modern neurosciences. *Neuropsychoanalysis.* https://doi.org/10.1080/15294145.2023.2284697

Copjec, J. (2015). *Read my desire: Lacan against the historicists* (2nd ed.). Verso.

Dauphin, B. (2023). Precursors of the affective neuroscience project in the writings of Melanie Klein. *Neuropsychoanalysis, 25*(2), 203–216. https://doi.org/10.1080/15294145.2023.2243280

Fotopoulou, A., & Tsakiris, M. (2017). Mentalizing homeostasis: The social origins of interoceptive inference. *Neuropsychoanalysis, 19*(1), 3–28. https://doi.org/10.1080/15294145.2017.1294031

Friston, K. (2010). The free-energy principle: A unified brain theory? *Nature, 11*(2), 127–138. https://doi.org/10.1038/nrn2787

Friston, K. (2017). Self-evidencing babies: Commentary on "Mentalizing homeostasis: The social origins of interoceptive inference" by Fotopoulou & Tsakiris. *Neuropsychoanalysis, 19*(1), 43–47. https://doi.org/10.1080/15294145.2017.1295216

Friston, K., & Frith, C. (2015). Active inference, communication and hermeneutics. *Cortex, 68*, 129–143. https://doi.org/10.1016/j.cortex.2015.03.025

Friston, K., Parr, T., Heins, C., Constant, A., Friedman, D., Isomura, T., Fields, C., Verblen, T., Ramsted, M., Clippinger, J., & Frith, C. (2024). Federated inference and belief sharing. *Neuroscience & Biobehavioral Reviews, 156*. https://doi.org/10.5500/j.neubiorev.2023.105500

Henderson, S. (2023). Defense mechanisms: A guide to brain functioning? *Neuropsychoanalysis, 25*(2), 191–202. https://doi.org/10.1080/15294145.2023.2261458

Hoşgören-Alıcı, Y., Hasanlı, J., Gradwohl, G., Turnbull, O., & Çakmak, E. (2023). Defense styles form the perspective of affective neuroscience. *Neuropsychoanalysis, 25*(2), 181–189. https://doi.org/10.1080/15294145.2023.2257718

Kernberg, O. (2022). Some implications of new developments in neurobiology for psychoanalytic object relations theory. *Neuropsychoanalysis, 24*(1), 3–12. https://doi.org/10.1080/15294145.2021.1995609

Lacan, J. (1954–1955/1991). *The seminar of Jacques Lacan, Book II: The ego in Freud's theory and in the technique of psychoanalysis* (J.-A. Miller, Ed.; S. Tomaselli, Trans.). Norton.

Lacan, J. (1964/1978). *The seminar of Jacques Lacan, Book XI: The four fundamental concepts of psychoanalysis* (J.-A. Miller, Ed., A. Sheridan, Trans.). Norton.

Leader, D. (2021). *Jouissance: Sexuality, suffering and satisfaction*. Polity.

Salas, C., Turnbull, O., & Solms, M. (Eds.). (2021). *Clinical studies in neurospychoanalysis revisited*. Routledge.

Solms, M. (2017). Some innate predictions are social in nature: Commentary on "Mentalizing homeostasis" by Fotopoulou and Tsakiris. *Neuropsychoanalysis*, *19*(1), 55–57. https://doi.org/10.1080/15294145.2017.1309622

Solms, M. (2020a). New project for a scientific psychology: General scheme. *Neuropsychoanalysis*, *22*(1–2), 5–35. https://doi.org/10.1080/15294145.2020.1833361

Solms, M. (2020b). Response to the commentaries on the "New Project." *Neuropsychoanalysis*, *22*(1–2), 97–107. https://doi.org/10.1080/15294145.2020.1843215

Solms, M. (2021). *Response to Otto Kernberg*. *Neuropsychoanalysis*, *23*(2), 115–119. https://doi.org/10.1080/15294145.2021.1984284

Žižek, S. (2020). *Sex and the failed absolute*. Bloomsbury.

Zupančič, A. (2017). *What is sex?* MIT Press.

12

Affects like Signifiers

Abstract Psychoanalysis often speaks of the power of putting feelings into words. However, this notion retains a differentiation between language and emotion. I propose that affects not only push to be connected to signifiers; affects themselves are organized like signifiers. Panksepp's basic emotional systems are structured like a language. Here, I demonstrate how affects can undergo linguistic-symbolic operations like displacement, condensation, and substitution through the prioritization function of the midbrain decision triangle. I further propose that the fundamental fantasy (as an abstract cognitive control hierarchy) organizes these chains of affect through predictions of precision.

Keywords Fundamental fantasy • Lacan • Neuroscience • Neuropsychoanalysis • Free energy principle • Predictive coding • Jouissance • Language

Let us pause for a brief clinical illustration. Mr. B presented with difficulty controlling pervasive PANIC, even when with fully supportive and caring partners. As we spoke, it became clear that, in at least some cases, the PANIC served a FEAR-function: express separation-distress to solicit a CARing partner, rather than an enRAGEd partner. In moments when we could discuss his associations to the sequence of events and thoughts leading up to PANIC-outbursts, it became clear that other affects sometimes preceded PANIC. In some cases, there was a LUSTful attraction which he found taboo and threatening to his attachment. In other instances, he spoke about deep RAGE that was too dangerous to discuss with his partner. Moreover, he complained that situations which used to be spaces for PLAY now left him wallowing in PANIC.

Affects like (Being Put into) Signifiers

A truism of psychoanalysis is that there is utility in putting affects into speech, finding "the words to say it," as Marie Cardinal (1975) elegantly put it. For Freud (1915a, 1915b), therapeutic work follows the reverse-direction of repression. Where repression disconnects the affect from the idea (with ensuing displacement onto a second idea or conversion into a bodily symptom), psychoanalysis aims to reconnect the affect to the original idea that was repressed. This process can remove (neurotic) symptoms via the interpretation of verbal associations—as illustrated in the study of dreams, jokes, slips, and so on.

These ideas undergird the Lacanian technical focus on speech, insofar as the play of signifiers undergirds the manifest thoughts and emotions that plague the patient. For Lacan, signifiers are *causal* regarding particular emotions, insofar as intervention through language can change affect, and *binding*, insofar as speech alleviates the intensity of affects and allows one to manage them (Busiol, 2021; Fink, 2011; Lacan, 1954–1955; Lacan, 1959–1960; Israely, 2018; Soler, 2015).

In computational terms, speech and language are domains of prediction (see Chap. 9), and affects are prioritized prediction errors (see Chap. 6). Insofar as predictions reduce prediction error, Solmsian neuropsychoanalysis could come to a similar conclusion as Lacan (except, perhaps, regarding the causality of language). It is good to put affects into words because that process itself deals with and contextualizes prediction error.

For Solms, declarative speech is a means of dealing with affects resulting from automatized (repressed) predictions:

> repression leads to endless, mindless *repetition*; which is why "transference" is so important in psychoanalytic treatment. Patients cannot re-think the repressed (since non-declarative memories cannot be retrieved into working memory), but they can think about what they are doing now, *in consequence* of the repressed. What patients *can* think about—i.e., can re-problematize, if it is brought to their attention—are the repetitive *derivatives* of the repressed, which involve *cortical* representations (of current experiences), which can therefore enter working memory and declarative (and reflective; i.e., prefrontal) thinking. This in turn allows their (derivative) predictions to be *reconnected with the affects that belong to them*, which enables the ego to *come up with better predictions*, with more realistic action plans, with the help of an adult brain (and that of the analyst) in adult circumstances. (Solms, 2018, p. 10, emphases in the original)

We can thereby say that *affects like (being put into) signifiers*. That is, affects "like" being put into words, insofar as that reduces uncertainty, problematizes habitual patterns (see Chap. 14), and facilitates dealing with prediction error. Words are especially helpful because they are the "mental solids" that form the basis for *thinking* through emotions and our strategies for managing them.

This was evident in Mr. B's treatment. We could only chart out the nuances in his emotional experience through the process of speech. Indeed, when we began treatment, he simply described "total anxiety" with no thoughts (other than separation) and physiological anxiety symptoms (heart racing, feeling tense). Slowing down and speaking allowed nuance and created space to better inhibit and control the waves of separation-distress.

Affects (Operate) like Signifiers

The notion that *affects like (being put into) signifiers* poses a contrast between "affect" (prioritized prediction error) and "signifier" (prediction). This follows the classical dichotomy in neuroscience and psychology between cognition and emotion (for review and criticism, see Cesario et al., 2020; Steffen et al., 2022). It implies two separate systems, one contained by the other.

However, the case of Mr. B indicates greater complexity. Alone, PANIC could be historicized to previous scenes of separation, rejection, traumatic relationships, and so on through speech. But putting PANIC into words revealed that there was more than just PANIC at play. Sometimes LUST preceded PANIC, sometimes it was RAGE. In other words, his affects were experienced in a *sequence* or *chain*. Sometimes Mr. B felt incredible guilt for his feelings (guilt is a blend of PANIC and RAGE, see below). That is, his feelings were sometimes (experientially) combined. Other times, a relatively positive situation—one which previously roused PLAY for Mr. B—instead aroused PANIC. One might say that PANIC was substituted for PLAY.

These are all operations—contiguous chaining, combination, substitution—that Lacan uses to characterize signifiers and the dynamics of the symbolic. These are Lacan's symbolic axes of metonymy and metaphor. Metonymy refers to the contiguous ordering of signifiers where displacement and condensation can occur. Metaphor refers to the substitutive selection of signifiers from a set (Dall'Aglio, 2023; Lacan, 1955–1956). However, in the present neuropsychoanalytic model, these operations—or something analogous to these operations—can occur at the level of *affect*.

Basic Emotions are Structured like a Language

Recall that Solms's (2021c) model of consciousness requires the existence of qualitatively distinct categories of homeostatic demand. Here, Panksepp's emotional systems are described as opposed to each other: SEEKING cannot be converted into RAGE. There is no common denominator or common currency to equalize different affects. Note that

12 Affects like Signifiers   175

this view of affect differs from Lacan's, where anxiety (as the common guise of *jouissance*) is a common currency for all emotion (Fink, 2011; Soler, 2015). In Solms's computational model, affects exist in a *differential system*. The oppositionality between affects is necessary for affective consciousness (see Chap. 8).

For Lacan, the symbolic is first and foremost a differential system (Lacan, 1954–1955, 1972–1973; Johnston, 2005). Importantly, this is not restricted to language. Lacan is precise on this point:

> I say that the unconscious is structured *like* a language. I say *like* so as not to say—and I come back to this all the time—that the unconscious is structured *by* a language. The unconscious is structured like the assemblages in question in set theory, which are like letters. (Lacan, 1972–1973, p. 48, emphasis in the original)

The unconscious is structured *like* a language—that is, structured by differential elements, representatives. Freudian *Vorstellungen* include word-presentations (*Wortvorstellungen*) and thing-presentations (*Sachevorstellungen*) (Freud, 1915b). *Vorstellungen*—Freud's ideational representatives—become Lacan's (1959–1960, 1964) signifiers. Images operate like signifiers when they are taken to be differential and enigmatic (Johnston, 2005). This is the work of piecemeal dream association, dream images as rebuses, and so on. Treating the image *like a word* (by putting it into speech) is possible—I suggest—because the image already has the potential to operate differentially. It is possible that this potential finds a neural parallel in the symbolic-imaginary knottings in visuospatial and linguistic perceptual processes (see Chap. 9).

With this Lacanian neuropsychoanalytic model, I claim that *affects (operate) like signifiers*. I say "like" to clarify that I am not equating affects to signifiers. However, symbolic operations typically attributed to signifiers also operate regarding affects (as demonstrated below). Indeed, the bedrock of affective consciousness is, for Solms, a differential system of hyperprior (and non-declarative) *predictions*. When one mines down to the Solmsian hidden spring of consciousness, one does not simply find pure emotion. One finds predictions. This fits with the Lacanian point that *there is no real outside the symbolic*, where pure emotion would be some non-symbolic substance that is captured, bound, and assuaged by speech. Rather, *there is only the symbolic*, but it is a symbolic that is *inconsistent* and *antagonistic*. Affective consciousness (the real of *jouissance*) emerges as an effect of structurally conflicting affective hyperpriors (an evolutionarily antagonistic differential system).

Indeed, this distinction between symbolic operations in language and symbolic operations in affects might shed light on Freud's (1925) insight that knowledge of the repressed does not undo the process of repression. Discovering the repressed idea and re-attaching its affect does not always remove the symptom, as many clinicians know. Likewise, clever Lacanian wordplay on a signifier does not always magically impact the patient. Linguistic interpretation hits a limit (Miller, 2023). I will return to these clinical implications in Chap. 14. Here I will describe how symbolic operations like displacement, condensation, and substitution can be conceptualized within the Lacanian neuropsychoanalytic model of basic emotions.

Displacement

For Lacan, displacement operates on the metonymic axis of the contiguous links among representations (Dall'Aglio, 2023). A cathexis can be displaced onto an associated representation. Traditional displacement speaks of affect between displaced from one idea to other (Freud, 1900). A dream element might be rendered unimportant compared to the rest of the narrative through displacement, for instance. But how can displacement (or something like it) be conceived for affects themselves?

Here, the distinction between J (the logical moment of uncertainty of uncertainty) and J(E) (prioritizing a specific category of uncertainty) is illuminating. Faced with uncertainty of uncertainty (J), the prioritization function selects one system: $J(E_1)$. Then this prioritization may shift to a second system: $J(E_2)$. There is an *informatic displacement*—not of an affect but of the prioritized system of uncertainty.

Consider cascade from PANIC to depressive downregulation of SEEKING (Panksepp, 1998; Solms & Turnbull, 2002). PANIC triggers separation distress and a protest-phase to SEEK out the lost caregiver.

Prolonged activation of PANIC triggers a dynorphin-mediated shutdown of the dopaminergic SEEKING system and a corresponding depressive phase (where low SEEKING activity corresponds to depressive affect). There is a chain from PANIC and high SEEKING to low SEEKING. Mr. B's experience offers another illustration of how affects can be *prioritized* in a sequence or chain: LUST [J(E_1)] followed by PANIC [J(E_2)], or RAGE [J(E_1)] then PANIC [J(E_2)]. To the extent that such a sequence of prioritizations is patterned or learned, emotions can exist in a contiguous structure.

Condensation

Freud (1900) describes condensation as the coming together of separate ideas, such as the superimposition of two images in a dream. Condensation can result from displacements of intensity, insofar as the condensed representation takes on the collective weight of the elements involved. For Lacan (1955–1956), condensation also operates on the metonymic axis (contrary to common Lacanian equations of condensation with metaphor, an equation Lacan himself sometimes makes). Insofar as condensation results from the combination of adjacent signifiers, it relies on a contiguous (i.e., metonymic) organization (Dall'Aglio, 2023).

Solms discusses the difference between basic emotions and secondary emotions that result from the combination of these emotions. PANIC combined with RAGE creates *guilt*; RAGE combined with FEAR creates *paranoia*; PLAY combined with RAGE creates *shame* (Lee & Solms, 2023). Also, recall the confluence of SEEKING (see Chap. 7). SEEKING blends with other affects to facilitate engagement with the external world. Moreover, PLAY involves the capacity to explore all the other emotional systems in an as-if fashion. Here, PLAY combines with other affects. PLAYing a game that involves FEAR and RAGE differs from pure FEAR or pure RAGE. Nor is the feeling simply PLAY. Neuropsychoanalysis already speaks of subjective experience (and not just signifiers) as itself a condensation of different feelings (see Smith & Solms, 2018).

Condensation of affects is possible through prediction, specifically learning how to deal with seven drives that are not only conflictual but also interactive:

> Consider sustainable romantic partnerships, for example, which require a judicious integration of LUST with childlike PANIC/GRIEF-type attachment (think of the Madonna-whore syndrome), which in turn is difficult to reconcile with the roving SEEKING drive (think of the thrill of novelty), as well as the inevitable frustrations that provoke RAGE (hence the ubiquity of domestic strife), which in turn conflicts with the concerns of nurturant CARE, and so on. Sustaining long-term relationships is just one example of the many challenges that face every human being. To manage these things—to manage life's problems—we use feelings as our compass. It is feeling that guides all learning from experience. (Solms, 2021a, p. 572)

Such combinations are possible through increasingly complex predictions of precision that prioritize *blends* of affects. Using our notation, we can write this as $J(E_1+E_2)$. A single feeling is *felt* (e.g., guilt) that is a combination of multiple trends (minimally, RAGE and PANIC). This is a condensation at the level of *feeling*—separate from that of the signifier.

Substitution

For Lacan (1955–1956, 1956–1957), substitution is the metaphoric process of replacing one signifier with another to produce a new signification. Novel signification is possible because metaphoric substitution involves rips a signifier from its pre-established lexical associations and quilts it to a new signified (Bazan et al., 2021; Dall'Aglio, 2023). Minimally, it is a selection process as opposed to a metonymic combinatorial process (Jakobson, 1956; Lacan, 1955–1956).

Recall that the prioritization function of affective consciousness takes place in the empty space of the subject. Prioritization of an emotional system does not follow because of a predictable context; it operates in the space of uncertainty (hence my retaining the notation of $J(E)$ to underline the centrality of uncertainty and aberrancy). Once again, this is because precision must itself be predicted. J must be prioritized as $J(E)$.

This creates a paradoxical logic. Recall Mr. A. He did not encounter the snake as a FEARful situation and then prioritize FEAR. The situation was not *in itself* FEARful. Rather, he encountered the snake (in a moment of uncertainty of uncertainty, J) and prioritized the situation via FEAR [J(FEAR)]. Now the situation is cast as a FEARful situation, insofar as *feeling* FEAR palpates through the hierarchy to adjust (the precision of) actions and perceptions in accordance with FEAR. FEAR prioritization renders the object a FEAR-object (Solms, 2021b).

Likewise for Mr. A's marital strife. The recommendation "do contemplative practice" is not frustrating (i.e., RAGE-inducing), leading Mr. A to *then* prioritize RAGE. Rather, the situation of hearing "do yoga" is encountered (J), and RAGE is inferred [J(RAGE)] to afford the greatest opportunity to minimize uncertainty. Then Mr. A's partner becomes an object of RAGE.

Or, recall Mr. B's complaint of experiencing situations that used to rouse social joy (PLAY) now as threats of separation-distress (PANIC). The situation does not determine the affect. Predictions of precision determine the prioritization function.

I claim that the prioritization function itself is a metaphoric operation, in the sense of metaphor as selection. A single affect from a set must be selected and others not. Moreover, a context that once led to prioritizing PLAY might subsequently arouse PANIC. This differs from PLAY *followed by* PANIC (displacement). A context previously predicted to be a good place to PLAY is now predicted to be cause for PANIC. The context has not changed, but the prediction of precision has. In other words, PANIC has been substituted for PLAY, producing a new (affective) meaning with respect to the object.

The Fundamental Fantasy Includes Predictions of Precision

These illustrations of *affects (operating) like signifiers* revolve around the prioritization function of affective consciousness. Recall, again, that this is a predictive process: predictions of precision. I have repeated this phrase several times. Now, it is helpful to consider one of Friston's expanded commentaries on it:

> The key thing here is to note that any inference rests upon an estimate of the (first-order) prediction error and the (second-order) precision of that error. In short, *two things have to be estimated* to make an inference—and this holds true for the brain. The biophysical realization of this dual estimation problem is in terms of the errors (e) and precisions (ω) that have distinct neuronal instantiations. Returning to the deep or hierarchal deployment of prediction errors in the brain, one might note that Solms emphasizes a distinction between interoception—that underwrites the exigencies of life—and exteroception. For one to inform the other, during inference, it is necessary that the "complexity of the interior of the organism, the nervous system" has a hierarchal structure apt for predicting both interoceptive and exteroceptive sensations. In other words, one can conceive of domain-specific sensory levels of a hierarchy upon which a domain-general level supervenes. This deep structure means that predictions about interoception can inherit beliefs based upon exteroceptive prediction errors and *vice versa*. On this reading, Solms' point is that the interoceptive prediction errors are assigned greater precision than exteroceptive prediction errors, by virtue of the fact that excursions from our homoeostatic setpoints are existentially more problematic than failures of exteroceptive prediction. Having said this, because the free energy pools prediction errors over the entire hierarchy, some interoceptive (homoeostatic) prediction errors can be tolerated if their cost is offset by a reduction of deep (domain general) prediction errors. (Friston, 2020, p. 58, emphasis in the original)

Friston here discusses how interoceptive and exteroceptive predictions can be entwined despite belonging to different domains with (generally speaking) different precisions. Interoceptive predictions are more precise than exteroceptive predictions due to homeostatic viable bounds. Entwinement occurs as domain-specific sensory areas are subsumed under a domain-general predictive level. To use a simple example, one can tolerate the pain of holding a hot cup of coffee over a short timescale (prediction errors relating to feeling pain) by prioritizing actions to hold the cup for a short timescale and bring it to a table (prediction errors relating to a goal-state). Given calculations of domain-general free energy, one category of prediction error is rendered more precise than another. This drives the action sequence and the selection of the predictive cascade "hold cup" over "drop painful object."

To prevent misunderstanding, the subsumption of domain-specific errors to domain-general free energy should not be read analogously to Freud's subsumption of partial drives under genital sexuality. This would return us to the beginning of our problems (see Chap. 4)! A domain-general calculation *does not do away with conflict* between domain-specific error categories. This holds true at the level of categorical emotions (for Solms) and the schematic effects of motor-inhibition (for Bazan). Nevertheless, the computational logic of domain-general free energy allows the flexibility of predictions of precision to lead to context-dependent prioritization.

I suggest that Friston's comments parallel the logical moment of J and J(E) at the level of differential basic emotions. The latter is the "complexity in the interior of the organism" highlighted by Solms (2020a) and commented upon above by Friston. In the face of uncertainty of uncertainty, different domain-specific categories of uncertainty (basic emotions) are considered (in a domain-general fashion), and one is prioritized. Or one is prioritized, then another (displacement). Or two are condensed together (combination). Or a situation that previously roused one system now leads to prioritizing another (substitution).

But what governs the structure of this domain-general level? What *organizes* the predictions of precision that guide the prioritization function which then leads to the operations of basic emotions like a language? I propose that the fundamental fantasy (as an abstract higher-order control rule in which premature automatized predictions are nested) *includes* predictions of precision of basic emotional systems. Stated otherwise, *the fundamental fantasy guides the prioritization function.*

Recall that the fundamental fantasy knots the real of *jouissance* to the symbolic and realizes its declarative (imaginary) implementation (see Chap. 9). It casts a certain stance between the subject and the Other and contextualizes uncertainty in the predictive field (*objet a*). This involves a *mode of relation* characterized by constellations and patterns of emotional stances toward the Other—in other words, patterns of how *jouissance* is prioritized in particular emotional systems [J(E)] in relation to the Other.

Recall the fantasy *My father is beating me* described by Freud. We can speculate on the network of Pankseppian emotions involved here, minimally: masochistic enjoyment (LUST condensed with pain), perhaps FEAR, and potential RAGE inferred in the Other. When displaced onto the father beating other children, the subject's enjoyment might turn sadistic (condensing LUST with RAGE). Disgust at the first hint at this fantasy by the analyst—epitomized in the subject's rejection of the possibility of this fantasy—might later be accepted, problematized, and joked about, disgust then being substituted for PLAY. Similarly, Mr. B's persistent prioritization of PANIC can be indexed to a fantasy *The Other is rejecting me* that undergirds much of their experience.

To the degree that an individual takes a general emotional stance toward the Other, one can say that predictions of precision are involved in the fantasy. The neurofunctional architecture for the fundamental fantasy—involving basal ganglia-prefrontal cortex loops whose abstract "blueprints" extend over sensorimotor regions—is well-suited to include predictions of precision *because it refers to a rule-hierarchy*. Such a *hierarchy* of predictions ranges from the most abstract of cognitive control rules, to ways of organizing perceptual information in association cortex, to learning reward predictive mappings, ultimately to the final motor output. The latter is the last point in the cycle for Solms's midbrain decision triangle: the midbrain locomotor region. Abstract control rules would involve predictions of precision at all these levels because predictions of input and predictions of precision are necessary for inference (Friston, 2020). To the extent that these predictions of precision converge on a common constellation of emotional systems in relation to the Other, one can speak of the fundamental fantasy at a deep level of prediction.

This fits with Friston's suggestion to complicate Solms's centripetal hierarchical model, where the midbrain decision triangle sits at the apex of the predictive hierarchy. Friston notes:

> Solms proposes that the ω-system [midbrain decision triangle] occupies a position deep in the hierarchy and is therefore "*resistant*" to belief-updating by prediction errors–a resistance endowed by learning over phylogenetic or ontogenetic timescales. But is it architecturally correct to place diencephalic

structures at the core of a centripetal hierarchy? In the sense that they broadcast to many hierarchical levels (and sensory modalities) this seems eminently sensible. On the other hand, these systems are in receipt of descending predictions (of precision) from many cortical areas and, implicitly, hierarchical levels. One might wonder whether a simple (centripetal) hierarchical architecture is sufficient to capture the pre-eminent role of the ω-system. (Friston, 2020, p. 60, emphasis in the original)

Importantly, Solms (2020b) shares Friston's hesitance of a simple centripetal (onion-like) hierarchy:

On the other hand, I do not mean to claim that "a simple (centripetal) hierarchical architecture is sufficient to capture the pre-eminent role of the ω-system"…What I have in mind is a *circular* causality, whereby centrifugal predictions with attendant precisions generate both stereotyped and voluntary actions, the sensory outcomes of the voluntary ones of which yield error signals whose actual precisions must be palpated in relation to the expected ones, yielding centripetal transmission of the residual errors arising from both classes of action–in relation to the originally expected free energy–leading to revised midbrain decisions (i.e. new precision assignments) and a sub-sequent cycle of centrifugal predictions. (Solms, 2020b, p. 102, emphasis in the original)

The fundamental fantasy, embedded in hierarchical control rules, would describe the organization of top-down predictive hierarchies and the "descending predictions (of precision)" that are integrated into the prioritization function of the midbrain decision triangle. It would be the higher-order counter-weight to the midbrain decision triangle in Solms's circular causality. Thus, the generalized mode of relation to the Other involves the pattern of prioritizing emotional systems between subject and Other, the pattern of how affects are organized like signifiers in the complex interpersonal human world (i.e., the predictive field).

This view of the fundamental fantasy fits with Solms's (2021a) approach to the Oedipus complex. For Solms, the Oedipus complex is not an innate primal fantasy but rather the inevitable outcome of the conflicts among Pankseppian emotions in the human interpersonal world. Solms privileges PLAY because it involves the capacity to explore all other affects in a safe manner *in relation to the needs and feelings of others*. Simply put, one cannot PLAY without an other—the reciprocality of the PLAY hyperprior necessitates this (Panksepp, 1998; Solms, 2021a). PLAY facilitates the resolution of the Oedipus complex, insofar as PLAY involves learning *how* to meet my conflicting needs in relation to others. It involves learning rules:

> It is easy to see how PLAY, in particular, gives rise to social rules. Rules regulate group behavior and thereby protect us from the excesses of our egocentric drives. It is also easy to see how social rules encourage complex forms of communication, and therefore how they contribute to the emergence of language and symbolic thought in general. The as-if quality of play suggests that it might even be the biological precursor of fantasy and thinking as a whole (i.e., of virtual versus real action). (Solms, 2021a, p. 576)

In other words, PLAY facilitates the resolution of the Oedipus complex, in part, by establishing rules and fantasy.

Where fantasy is preconscious for Solms because it involves declarative representations (Smith & Solms, 2018), for Lacan the fundamental fantasy is an unconscious structure. I propose that the unconscious fantasy emerges from the rules that PLAY gives rise to in relation to the Oedipal conflict. Specifically, the fundamental fantasy could be described as the (symbolic) rules for mediating enjoyment in relation to the Other. Indeed, for Lacan, the dissolution of the Oedipus complex involves the installation of the fundamental fantasy (Lacan, 1955–1956, 1956–1957). Insofar as the contradictions of affective hyperpriors are indexed to the interpersonal predictive field (where certain figures occupy the place of the Other), the fantasy is the answer to the question *how do I deal with my contradicting affects in relation to the Other?*

References

Bazan, A., Van de Vijver, G., & Caine, D. (2021). Lacanian neuropsychoanalysis: On the role of language motor dynamics for language processing and for mental constitution. In C. Salas, O. Turnbull, & M. Solms (Eds.), *Clinical studies in Neuropsychoanalysis revisited* (pp. 79–104). Routledge.
Busiol, D. (Ed.). (2021). *Lacanian psychoanalysis in practice: Insights from fourteen psychoanalysts*. Routledge.
Cardinal, M. (1975/2003). *The words to say it* (P. Goodheart, Trans.). Van-Vactor & Goodheart.
Cesario, J., Johnston, D., & Eisthen, H. (2020). Your brain is not an onion with a tiny reptile inside. *Current Directions in Psychological Science, 29*(3), 255–260. https://doi.org/10.1177/0963721420917687
Dall'Aglio, J. (2023). Lacan's use of neurology: A neuropsychoanalytic reading of Seminar III, lessons XVII and XVIII. *Lacunae, 25*, 26–73.
Fink, B. (2011). *Fundamentals of psychoanalytic technique: A Lacanian approach for practitioners*. Norton.
Freud, S. (1900/2010). *The interpretation of dreams* (J. Strachey, Ed. & Trans.). Basic Books.
Freud, S. (1915a/1957). Repression. In *The standard edition of the complete psychological works of Sigmund Freud, Vol. XIV* (J. Strachey, Ed., Trans.) (pp. 141–158). Hogarth Press.
Freud, S. (1915b/1957). The unconscious. In *The standard edition of the complete psychological works of Sigmund Freud, Vol. XIV* (J. Strachey, Ed., Trans.) (pp. 159–215). Hogarth Press.
Freud, S. (1925/1961). Negation. In *The standard edition of the complete psychological works of Sigmund Freud, Vol. XIX* (J. Strachey, Ed., Trans.) (pp. 233–240). Hogarth Press.
Friston, K. (2020). The importance of being precise: Commentary on "New project for a scientific psychology: General scheme" by Mark Solms. *Neuropsychoanalysis, 22*(1–2), 57–61. https://doi.org/10.1080/15294145.2021.1878610
Israely, Y. (2018). *Lacanian treatment: Psychoanalysis for clinicians*. Routledge.
Jakobson, R. (1956/1990). Two aspects of language and two types of aphasic disturbances. In L.R. Waugh & M. Monville-Burston (Eds.) *Roman Jakobson: On language* (pp. 115–133). Harvard University Press.
Johnston, A. (2005). *Time driven: Metapsychology and the splitting of the drive*. Northwestern University Press.

Lacan, J. (1954–1955/1991). *The seminar of Jacques Lacan, Book II: The ego in Freud's theory and in the technique of psychoanalysis* (J.-A. Miller, Ed.; S. Tomaselli, Trans.). Norton.

Lacan, J. (1955–1956/1997). *The seminar of Jacques Lacan, Book III: The psychoses* (J.-A. Miller, Ed., R. Grigg, Trans.). Norton.

Lacan, J. (1956–1957/2021). *The seminar of Jacques Lacan, Book IV: The object relation* (J.-A. Miller, ed., A. R. Price, Trans.). Polity.

Lacan, J. (1959–1960/1992). *The seminar of Jacques Lacan, Book VII: The ethics of psychoanalysis* (J.-A. Miller, Ed., & D. Porter, Trans.) Norton.

Lacan, J. (1964/1978). *The seminar of Jacques Lacan, Book XI: The four fundamental concepts of psychoanalysis* (J.-A. Miller, Ed., A. Sheridan, Trans.). Norton.

Lacan, J. (1972–1973/2000). *The seminar of Jacques Lacan, Book XX: On feminine sexuality, the limits of love and knowledge* (J.-A. Miller, Ed., B. Fink, Trans.). Norton.

Lee, T., & Solms, M. (2023). Managing the clinical encounter with patients with personality disorder in a general psychiatry setting: Key contributions from neuropsychoanalysis. *BJPsych Advances*, 1–8. https://doi.org/10.1192/bja.2023.43

Miller, J.-A. (2023). *Analysis laid bare*. Libretto Press.

Panksepp, J. (1998). *Affective neuroscience: The foundations of human and animal emotions*. Oxford University Press.

Smith, R., & Solms, M. (2018). Examination of the hypothesis that repression is premature automatization: A psychoanalytic case report and discussion. *Neuropsychoanalysis*, 20(1), 47–61. https://doi.org/10.1080/15294145.2018.1473045

Soler, C. (2015). *Lacanian affects: The function of affect in Lacan's work* (B. Fink, Trans.). Routledge.

Solms, M. (2018). The neurobiological underpinnings of psychoanalytic theory and therapy. *Frontiers in Behavioral Neuroscience*, 12, 294. https://doi.org/10.3389/fnbeh.2018.00294

Solms, M. (2020a). New project for a scientific psychology: General scheme. *Neuropsychoanalysis*, 22(1–2), 5–35. https://doi.org/10.1080/15294145.2020.1833361

Solms, M. (2020b). Response to the commentaries on the "New Project." *Neuropsychoanalysis*, 22(1–2), 97–107. https://doi.org/10.1080/15294145.2020.1843215

Solms, M. (2021a). A revision of Freud's theory of the biological origin of the Oedipus complex. *Psychoanalytic Quarterly, 90*(4), 555–581. https://doi.org/10.1080/00332828.2021.1984153

Solms, M. (2021b). Response to Otto Kernberg. *Neuropsychoanalysis, 23*(2), 115–119. https://doi.org/10.1080/15294145.2021.1984284

Solms, M. (2021c). *The hidden spring: A journey to the source of consciousness.* Profile Books.

Solms, M., & Turnbull, O. (2002). *The brain and the inner world: An introduction to the neuroscience of subjective experience.* Other Press.

Steffen, P., Hedges, D., & Matheson, R. (2022). The brain is adaptive not triune: How the brain responds to threat, challenge, and change. *Frontiers in Psychiatry, 13*, 802606. https://doi.org/10.3389/fpsyt.2022.802606

13

Toward Levels of the Symbolic

Abstract By linking the Lacanian symbolic to the brain, one finds that the logic of symbolic organization applies to more than just language *per se*. Lacan himself is clear to articulate a particular linguistic organization to the unconscious which is not equated with specific language. Hence, the unconscious is structured *like* a language. This chapter sketches *levels of the symbolic*—different domains of brain systems structured like a language—including basic emotions, the motor axis, language motor dynamics, motoric representations, and abstract cognitive control.

Keywords Language • Panksepp • Motor trace • Phonology • Lacan • Neuroscience • Neuropsychoanalysis • Affect • Emergence

Although affects operate like signifiers, they are surely qualitatively distinct from motoric signifiers used to manage prediction error. Moving up the predictive hierarchy, although abstract cognitive control rules are linguistically organized (propositional logic, conditionals, etc.), surely they are distinct from spoken language *per se*. Decades of neuropsychological research on the dissociation of mental faculties supports this (Lezak et al.,

2012), as does psychoanalytic work with brain injury patients (Kaplan-Solms & Solms, 2002; Salas et al., 2021). The recognition of domain-specific prediction errors likewise necessitates qualitatively distinct predictive levels (Friston, 2020).

If I characterize all these neuropsychological functions (affects, general motor functions, language motor dynamics, abstract control rules) as following the logic of the signifier, then I risk reduction of all of these systems to one Lacanian symbolic. In order to sustain the nuance demonstrated by neuropsychological research into hierarchical organization of the brain, then a Lacanian neuropsychoanalysis must admit *levels of the symbolic*.

A full discussion of different levels of the symbolic is beyond the scope here. For now, I will sketch these different levels and their place within a Lacanian meta-neuropsychology. First, consider Leader's reflections on bodily innervation that frequently get reduced to the "real of *jouissance*" in Lacanian theory:

> The supposed jouissance of the body must be a complex field prior to the supposed imposition or inscription of the symbolic, with its own rhythms, patterns of discharge, urgencies, relations to musculature, endocrine effects, links to respiration, and so on. (Leader, 2021, p. 104)

Leader's call for Lacanians to investigate biological processes points to a significant theoretical amendment. The "complex field" prior to the symbolic is not simply the vague negativity of the real. Nor is it a clearly predictable reflex arc. There are "patterns" of discharge, "rhythms," "endocrine effects," "relations to musculature," and so on. I suggest interpreting this "complex field" as a symbolic, differential system prior to the inscription of language. Here, I will sketch different levels of the symbolic that have been alluded to in previous chapters. In concert with Leader, I propose that these areas are fruitful openings for future interdisciplinary Lacanian research.

This brief discussion also serves a crucial theoretical point. Recall that transcendental materialism advocates for a view of "weak nature" rift with antagonisms (see Chap. 3). With increasing complexity and differentiality come short-circuits that produce phenomena irreducible to the prior level (Johnston, 2019). I suggest that the different levels of the symbolic be considered through the lens of emergence. Emergence of novel elements of subjectivity occur at points of antagonistic complexity in the organism.

Basic Emotions are Structured like a Language

The first level we can discern is that signifier-organization of basic emotions (see Chap. 12). Differential affective hyperpriors conflict, leading to the emergence of affective consciousness as surplus prediction error. Here, the psychoanalytic real of *jouissance* emerges from the deadlock of the neuronal real of antagonism between emotional instincts. Put otherwise, a unique phenomenon—affective consciousness as subjectively *felt uncertainty*—emerges as a short-circuit, the lack of a meta-prediction. Moreover, through predictions of precision in the fundamental fantasy, affects follow operations analogous to condensation, displacement, and substitution. There is no "true," "original" affect—one begins from contradiction.

The notion that affects operate like signifiers resonates with Lacan's characterization of the Freudian id which, for Solms, corresponds to Panksepp's emotional systems (see Chap. 6):

> If analysis has brought us something, it is the following. The Es [id] is not a physical reality, nor it is merely what was there before. The Es is organised and articulated as the signifier is organised and articulated. (Lacan, 1956–1957, p. 38)

For Lacan, there is something of the signifier *already there in material reality*:

> The Es that is at issue in analysis is the signifier that is already there, in the real, the uncomprehended signifier. It's already there, but it's a signifier, and not some confused and primitive property of goodness knows what pre-set harmony. (Lacan, 1956–1957, p. 41)

Lacan's issue with the *Es* as a "physical reality"—which, in our terms, should be restated as "physiological" insofar as physics differs from physiology (Solms, 2021)—is the "primitive property of goodness knows what pre-set harmony" that often accompanies views of nature. This would be the capital-N Nature of fully harmonious laws (see Chap. 3; Johnston, 2019). To say that there is a "signifier that is already there, in the real" designates material reality as differential and not harmonious. This would be the "weak nature" of cracks and short-circuits in transcendental materialism. More specifically, for Lacan, this signifier already in the real is the Freudian id.

But surely the *Es* as "uncomprehended signifier" differs from the signifiers of everyday speech—especially if one retains the link between drive and id. Nevertheless, Lacan claims that the id is "organised and articulated" like the signifier. Id-as-signifier thereby implies a different *level of the symbolic*.

Lacan alludes to something like a level of the symbolic prior to language proper in his discussion of certain features of the human body as "proto-signifiers":

> There are a number of elements, of accidental occurrences of the body, that amount to experiential givens. Just as in nature, certain natural reservoirs are already there, so too in the signified there are certain elements that are taken up in the signifier in order to furnish it with, as it were, its first weapons. These elements are things that are ungraspable in the extreme, and yet they are irreducible. Among them is the phallic term, the pure and simple erection…a number of elements are each linked more or less to bodily stature, and not merely to the lived experience of the body. They constitute the first elements, and are effectively borrowed from experience, but completely transformed by the fact that they are symbolised. Symbolised means that they are introduced into what characterises the bond of the signifier as such, the signifier being something that is articulated in keeping with its own logical laws. (Lacan, 1956–1957, p. 43)

Lacan's reliance on the imaginary to theorize the body (as acting and feeling)—especially in the mid-1950s—perhaps leads him astray to the *image* of phallic erection as a proto-signifier. He misses the differential nature of actions and emotions (non-declarative bodily processes) themselves.

I propose that the "accidental occurrences of the body" that relate to "bodily stature" are remarkable because they *stand out*—that is, they are *differentiated*—from other features of the body. Specifically, these are action patterns (non-declarative procedural and emotional predictions) that deal with bodily tension (Bazan & Detandt, 2013). They are "ungraspable in the extreme" because they point to predictions logically prior to semantic and linguistic grasp.

Furthermore, one need only connect Lacan's id-as-signifier to these proto-signifier elements to recognize that the Freudian id (Panksepp's basic emotional hyperpriors) is one such naturally-given accident of experience that is differential (not harmonious) in structure. To claim, as I do, that Pankseppian basic emotions are structured like a language, thus resonates with Lacan's claim that the id is "organised and articulated as the signifier is organised and articulated"—that is, by axes *like* metonymy and metaphor.

Importantly, I am not suggesting that affects operate like signifiers because we speak (e.g., Busiol, 2021). This is a reductive Lacanian argument (often made by Lacan himself) which prohibits any discussion of human dimensions outside of language or pre-linguistic phylogenetic history (Johnston, 2019). While the degree of complexity (e.g., turning RAGE against the self) likely depends on the degree of complexity in the predictive model, the level of basic emotions is already *structured like a language*.

A Motoric Symbolic

With vertebrate organisms (which extend beyond the mammals studied by Panksepp) another antagonism is introduced with the splitting off of the motor axis from bodily homeostasis (see Chap. 7). This introduces a minimal degree of active inference—that is, the intentional selection of actions that are predicted to minimize uncertainty—beyond innate reflex arcs and perceptual belief updating. This is another way of saying that *learning* must occur.

With only reflexes (e.g., the patellar tendon reflex: the doctor hits the tendon under your kneecap and your quadriceps contracts; no thinking or feeling is necessary), motoric predictions cannot be said to be symbolic. This is because there is a one-to-one relation between prediction error and prediction. The error signals triggered by the stretching of the patellar tendon trigger a prediction that is guaranteed to return the body to the expected sensory state. The same can be said of autonomic interoceptive predictions of which we cannot be conscious (e.g., predictions adjusting blood pressure to remain within homeostatic bounds). This is because their precision-weights are monotonous (Solms, 2021). Here, signifier is soldered to signified without any ambiguity.

When in the space of uncertainty—specifically that opened by affective consciousness as *felt uncertainty* (i.e., the adjustment of precision-weights)—the phenomenon of contingent, historicized learning emerges. For motor learning, this is the dopamine spike tagging mechanism that marks the motor prediction which surprisingly alleviates some drive tension. This contingency takes place in a rift between bodily need and motor axis. Here, a minimal repetition compulsion emerges in the form of incentive sensitization—that energized tendency to repeat the tagged trace independent of need-activation. This is due to the high precision assigned to the motor prediction through SEEKING (see Chap. 7).

At the level of general motor functions as signifiers, the signified is not a meaning. For motor predictions that are historicized to deal with drive-tension, the analogue to the "signified" would be the reduction of a particular category of homeostatic uncertainty. For example, the motor predictions involved in the PANIC cascade are quilted to the (hoped for outcome) of reunion with the caregiver, which corresponds to a reduction of PANIC prediction error. Importantly, such predictions are minimally symbolic insofar as they must be learned—that is, they are (1) not innate and (2) differentiated from other motor actions the organism could take. All of this occurs without semantic meaning, at least in the human sense—although it is likely that certain perceptual images are associated with the reduction of prediction error and therefore might be associated with the motor signifier prediction. Simply put, different levels of the symbolic implies that the quality of the signified differs accordingly.

Phonological Ambiguity in Language Motor Dynamics

With increasing complexity and the formation of shared generative models that require some means for communication between agents, a new level emerges: language (Friston et al., 2024). One could say that the antagonism here is between differentiated agents who attempt to form a shared generative model and perform federated inference. The gap between agents creates the pressure to create some method of belief-sharing. This leads to the emergence of symbolic communication systems to minimize free energy.

It is beyond the scope of this paper to differentiate animal communication from human language. For now, it suffices to recall that language is first and foremost a motor operation (Bazan, 2011). This highlights the symbolic dimension of language. For psychoanalysis, language does not primarily serve the function of communication. Speech is a system of fine motor articulatory patterns.

The novel cut introduced by language is phonological ambiguity with respect to semantic predictions. With linguistic motor predictions, identical (or similar) patterns can produce different meanings. To give a simple example, the same motor predictions to pronounce *sharp* can lead to different semantic inferences: intelligent, cutting, and so on. Conversely, very different motor patterns can converge on the same meaning. Language is a notably precarious motor system insofar as particular traces can take on entirely different meanings depending on their ordering and organization (Bazan, personal communication). Fine motor speech sequences require predictive parsing to extract semantic understanding (Bazan et al., 2021). The traditional Lacanian $ as divided by the signifier can be situated at this level.

Note that, with this level, we are bridging non-declarative predictions (emotional memories, procedural memories) and declarative predictions (semantic memory). There is an antagonism between semantic meaning and linguistic-motoric prediction. For me, this is why language is special: it bridges non-declarative and declarative predictive levels. Motoric signifiers for speech have signifieds in the semantic domain as well as motor dynamics of the dopamine spike tagging mechanism. This has important clinical implications which I develop in Chap. 14.

Motoric Representations and Abstract Cognition

The differential nature of motoric predictions and the capacity for linguistic organization leads to further complexity. Most simply, the need to inhibit some predictions and execute others (through precision-modulation) leaves non-attenuated tension in charged motor traces. Bazan (2011) argues that these *inhibited* motoric predictions a reworked into higher-order motor schemas that *unconsciously* impact decision-making. Bazan and colleagues have conducted several experiments demonstrating the independent impact of *inhibited* phonological ambiguity on choice, anxiety, and conscious interpretation (Bazan et al., 2019; Bruxelmane et al., 2020; Olyff & Bazan, 2023; Thieffry et al., 2023). Here is the symbolic unconscious typically conceived in Lacanian psychoanalysis. Notably, this symbolic unconscious (of words) differs from Solms's non-declarative procedural unconscious of prematurely automatized predictions. These are different symbolic levels.

With linguistic organization (of speech and motor schemas) comes a further level of complexity. Conditional and propositional logic organizes sequences of predictions. This is the essence of hierarchical cognitive control (see Chap. 9). Such organizational complexity does not only facilitate the reduction of prediction errors generated at lower levels. The logical tools introduced by language and syntactical structure result in *predictive systems that generate their own hidden states*. As Murphy et al. (2022) put it:

> Indeed, generating complex, structure recursive thoughts expands the range of possible hidden states, such that there are an increased number of states that the world could look like to the organism. Possibilities for uncertainty resolution rapidly expand with complex syntax. (Murphy et al., 2022, p. 33)

For example, linguistic conceptual organization introduces certain categories of objects. A system must make several inferences, then, regarding an object: color (black), line orientation (four sides at 90-degree angles), apperception (a phone), function (for calling, texting, internet scrolling), living or non-living (non-living), who it belongs to, and so on. Creation of concepts and categories through increasingly complex predictive structures increases the possibilities of how the world can be interpreted (i.e., expands the number of hidden states). In other words, new nodal points of the real emerge from complex predictive syntax. From a transcendental materialist perspective, one can put this formulaically: complexity breeds negativity.

Summary of Levels of the Symbolic

Additional discussion of these levels is beyond the scope of this book. For now, recall that Lacan (1955–1956)—following Jakobson (1956)—proposes that symbolic principles like metaphor and metonymy chart from the organization of phonemes to phrases, sentences, conversation, discourse, and culture writ large. I claim that, while this organizational isomorphism might be true, it is not license to collapse all of these levels onto a monolithic "symbolic." Different levels of the predictive hierarchy are qualitatively distinct and domain-specific. Metaphor at the sentence level differs from general meaning comprehension—as studies comparing micro- and macro-level aphasic discourse demonstrate (Dall'Aglio, 2023; Linnik et al., 2016; Marini et al., 2011). Likewise for consider metonymy. The metonymy of Schreber's delusional sentence fragments or in the speech of Wernicke's aphasics (Lacan, 1955–1956) differs from the metonymic contiguity of Little Hans's fantasies (Lacan, 1956–1957).

To be clear, I do not seek to downplay the significance of language—especially in psychoanalytic practice as an exchange of speech. However, symbolic systems can be discerned prior to the imposition of language, and abstract symbolic processes emerge that are perhaps dependent on linguistic organization (e.g., propositional logic) but are not identical to language *per se*. This fits with Lacan's later push toward topology and mathematics—a degree of abstraction that is *like* a language but not identical to language (Lacan, 1972–1973).

References

Bazan, A. (2011). Phantoms in the voice: A neuropsychoanalytic hypothesis on the structure of the unconscious. *Neuropsychoanalysis, 13*(2), 161–176. https://doi.org/10.1080/15294145.2011.10773672

Bazan, A., & Detandt, S. (2013). On the physiology of jouissance: Interpreting the mesolimbic dopaminergic reward functions from a psychoanalytic perspective. *Frontiers in Human Neuroscience, 7*, 709. https://doi.org/10.3389/fnhum.2013.00709

Bazan, A., Kushwaha, R., Winer, E. S., Snodgrass, J. M., Brakel, L., & Shevrin, H. (2019). Phonological ambiguity detection outside of consciousness and its defensive avoidance. *Frontiers in Human Neuroscience, 13*, 77. https://doi.org/10.3389/fnhum.2019.00077

Bazan, A., Van de Vijver, G., & Caine, D. (2021). Lacanian neuropsychoanalysis: On the role of language motor dynamics for language processing and for mental constitution. In C. Salas, O. Turnbull, & M. Solms (Eds.), *Clinical studies in Neuropsychoanalysis revisited* (pp. 79–104). Routledge.

Bruxelmane, J., Shin, J., Olyff, G., & Bazan, A. (2020). Eyes wide shut: Primary process opens up. *Frontiers in Psychology, 11*, 145. https://doi.org/10.3389/fpsyg.2020.00145

Busiol, D. (Ed.). (2021). *Lacanian psychoanalysis in practice: Insights from fourteen psychoanalysts*. Routledge.

Dall'Aglio, J. (2023). Lacan's use of neurology: A neuropsychoanalytic reading of Seminar III, lessons XVII and XVIII. *Lacunae, 25*, 26–73.

Friston, K. (2020). The importance of being precise: Commentary on "New project for a scientific psychology: General scheme" by Mark Solms. *Neuropsychoanalysis, 22*(1–2), 57–61. https://doi.org/10.1080/15294145.2021.1878610

Friston, K., Parr, T., Heins, C., Constant, A., Friedman, D., Isomura, T., Fields, C., Verblen, T., Ramsted, M., Clippinger, J., & Frith, C. (2024). Federated inference and belief sharing. *Neuroscience & Biobehavioral Reviews, 156*, 105500. https://doi.org/10.1016/j.neubiorev.2023.105500

Jakobson, R. (1956/1990). Two aspects of language and two types of aphasic disturbances. In L. R. Waugh & M. Monville-Burston (Eds.) *Roman Jakobson: On language* (pp. 115–133). Harvard University Press.

Johnston, A. (2019). *Prolegomena to any future materialism, Volume two: A weak nature alone*. Northwestern University Press.

Kaplan-Solms, K., & Solms, M. (2002). *Clinical studies in neuro-psychoanalysis: Introduction to a depth neuropsychology* (2nd ed.). Karnac Books.

Lacan, J. (1955–1956/1997). *The seminar of Jacques Lacan, Book III: The psychoses* (J.-A. Miller, Ed., R. Grigg, Trans.). Norton.

Lacan, J. (1956–1957/2021). *The seminar of Jacques Lacan, Book IV: The object relation* (J.-A. Miller, ed., A. R. Price, Trans.). Polity.

Lacan, J. (1972–1973/2000). *The seminar of Jacques Lacan, Book XX: On feminine sexuality, the limits of love and knowledge* (J.-A. Miller, Ed., B. Fink, Trans.). Norton.

Leader, D. (2021). *Jouissance: Sexuality, suffering and satisfaction*. Polity.

Lezak, M., Howieson, D., Bigler, E., & Tranel, D. (2012). *Neuropsychological assessment* (5th ed.). Oxford University Press.

Linnik, A., Bastiaanse, R., & Höhle, B. (2016). Discourse production in aphasia: A current review of theoretical and methodological challenges. *Aphasiology, 30*(7), 765–800. https://doi.org/10.1080/02687038.2015.1113489

Marini, A., Andreetta, S., del Tin, S., & Carlomagno, S. (2011). A multi-level approach to the analysis of narrative language in aphasia. *Aphasiology, 25*(11), 1372–1392. https://doi.org/10.1080/02687038.2011.584690

Murphy, E., Holmes, E., & Friston, K. (2022). Natural language syntax complies with the free-energy principle. *arXiv*. https://doi.org/10.48550/arXiv.2210.15098

Olyff, G., & Bazan, A. (2023). People solve rebuses unwittingly—Both forward and backward: Empirical evidence for the mental effectiveness of the signifier. *Frontiers in Human Neuroscience, 16*, 965183. https://doi.org/10.3389/fnhum.2022.965183

Salas, C., Turnbull, O., & Solms, M. (Eds.). (2021). *Clinical studies in neurospychoanalysis revisited*. Routledge.

Solms, M. (2021). *The hidden spring: A journey to the source of consciousness*. Profile Books.

Thieffry, L., Olyff, G., Pioda, L., Detandt, S., & Bazan, A. (2023). Running away from phonological ambiguity, we stumble upon our words: Laboratory induced slips show differences between highly and lowly defensive people. *Frontiers in Human Neuroscience, 17*, 1033671. https://doi.org/10.3389/fnhum.2023.103367

14

Clinical Lacanian Neuropsychoanalysis

Abstract This chapter applies a Lacanian lens to Solms's clinical neuropsychoanalytic model. It also formulates Lacanian technique (especially punctuation and scansion) in neuropsychoanalytic terms as provoking surprise. I describe how psychoanalytic (including Lacanian) intervention utilizes PLAY to modulate deeply automatized predictions. Additionally, I propose a model of Lacanian neuropsychoanalytic clinical formulation and intervention along the lines of symbolic reduction. Repressed predictions (S1s) may be doomed to repeat, but an encounter with the antagonism at the core of one's predictive models—predictive destitution—may open flexibility in *how* they repeat.

Keywords Scansion • Punctuation • Technique • Neuropsychoanalysis • Symbolic • Reduction • Play • Jouissance

In this penultimate chapter, I will discuss clinical contributions from this dialogue between Lacanian psychoanalysis, neuroscience, and

---

Here I develop arguments from Dall'Aglio (2021).

© The Author(s), under exclusive license to Springer Nature Switzerland AG 2024
J. Dall'Aglio, *A Lacanian Neuropsychoanalysis*, The Palgrave Lacan Series,
https://doi.org/10.1007/978-3-031-68831-7_14

neuropsychoanalysis. To be clear, these contributions are suggestions, ideas that may be more or less helpful to different clinicians and different patients. I hope this dialogue opens space to consider how these interdisciplinary concepts can stimulate new ways for thinking about case formulation and intervention.

Solmsian Clinical Neuropsychoanalysis

Solms's neuropsychoanlaytic formulation follows directly from his meta-neuropsychological model (Solms, 2018). Consciousness is fundamentally affective. This is what the patient is suffering from: conscious *feelings* of unpleasure. Patients consult professionals with a wish to remove the feeling: relieve anxiety, escape depression, overcome phobias, manage rage, feel closer to others, and so on. The first step is to identify the consciously felt affect. Recall Mr. B. His life was filled with pervasive separation distress. So many situations triggered unbearable PANIC. Or recall Mr. A: the encounter with the snake led to clear FEAR; feeling chronically misunderstood by his husband resulted in consistent underlying RAGE.

Next, one must identify the unconscious (non-declarative) prediction that is *causing* the conscious affect. Patients are by definition unaware of the unconscious prediction because it is non-declarative. It cannot be formulated in declarative terms in self-reflective working memory. Nevertheless, they do not work. Repetition of prematurely automatized predictions results in ongoing prediction error that is prioritized and felt as affective consciousness. This is how the repressed prediction causes the conscious affect.

Moreover, this is why Solms claims that the repressed does not return (into consciousness). The repressed (premature automatized prediction) remains non-declarative and unconscious. What returns are the *consciously felt* affects and the *declarative derivatives* of the repressed. Derivatives are echoes of the repressed prediction—often modified, supplemented, or disguised by secondary defensive predictions—in dreams, fantasies, interpersonal patterns, transference, symptoms, and so on. One *constructs* the repressed prediction in the declarative language of the ego through its refraction in different, derivative scenes.

Recall Mr. A's strategy to "separate." Faced with frustration at his husband's recommendation to "do contemplative practice," Mr. A considers responding with dissatisfaction but ultimately says nothing. One might say that "separate" is a prematurely automatized prediction that Mr. A uses to deal with conflict: "If I separate responding from my perspective, then I will avoid the other's anger." This prediction is a bad solution to the problem of RAGE. Yet, this is the hallmark of prematurely automatized predictions: they do not work.

Solms's neuropsychoanalytic technique follows in the reverse direction from manifest symptom-formation (i.e., derivatives and conscious affect). Interpretation aims to reconnect the declarative derivatives of the non-declarative repressed with the affects that belong to them. For example, one might point out to Mr. A how his strategy to "separate"—in this declarative example and other instances where the prediction repeats—contributes to his ongoing frustration.

However, because of the nature of premature automatized predictions, they are not subject to reconsolidation (modification) in self-reflective working memory. They will thereby continue to repeat. Solms argues that one must continue to point out and interpret this repetition in multiple derivative scenes so that the patient can gradually consolidate new, better predictions that better meet the emotional need. To be clear, Solms is not suggesting educating patients or teaching them better predictions. Psychoanalysis is a creative space to explore—in a slowed down time with greater resources than childhood—new ways of dealing with the exigencies of life. Solms calls for *problematizing* predictions, not teaching solutions.

This process is slow, reflecting the difficulty of working through. Moreover, consolidating new non-declarative action plans does not remove the prematurely automatized predictions, which may still be reactivated in times of stress. Hence the ongoing possibility of regression. Nevertheless, through repeated declarative work guided by an inference of the non-declarative prediction, new non-declarative predictions can slowly be formed.

Before continuing, it is helpful to explicitly state how Solms's approach differs from the "corrective emotional experience" perspective in contemporary psychotherapy (e.g., Castonguay & Hill, 2012; Levenson, 1995).

In short, this view argues that the therapist does not play into the patient's old relational, cognitive, or emotional habits, disconfirming them, and providing a "corrective" (emotional, behavioral, or cognitive) experience that allows the patient to feel, think, or act differently. Such an experience was lacking earlier in life, likely in the parental matrix. The new experience with the therapist challenges the patient's old expectations rather than confirming them.

I will not comment on the validity or utility of this approach. Solms's model differs, however, in that the role of the analyst is not to disconfirm the repressed prediction. Because such predictions are *prematurely* automatized, *they are constantly disconfirmed*. Hence, the patient suffers from unpleasant affect. The analyst does not disconfirm predictions. The analyst *problematizes* them, raising their derivatives to working memory to *think* about what is happening (Smith & Solms, 2018).

Lacanian Contributions to Neuropsychoanalytic Technique

Here I will discuss how Lacanian techniques can be situated in and contribute to clinical neuropsychoanalysis. I specifically have in mind techniques that focus on the signifier: punctuation and scansion (Fink, 2011; Israely, 2018). To begin with a general comparison, one might say that Solmsian neuropsychoanalytic technique focuses on tolerating prediction error and problematizing predictions. Lacanian technique extends this to *provoking* prediction error to *modulate* predictions.

Punctuation

Let us return to Mr. A. His husband's name was Ben. Mr. A and his husband had a close friend, also named Ben, to whom they would also vent their frustrations and turn for advice. This friend recommended that they both try to pause more. When in a conflict, he advised that they try to see where the other person was coming from. Mr. A took this advice seriously but found it frustrating. He felt that he was making a genuine effort to empathize and recognize that his husband's perspective was "separate" from his own. But he felt that his husband usually failed to do this.

One session, Mr. A spoke at length about a heated argument with his husband. Ben was "going off" about something that Mr. A wanted nothing to do with. This time, Mr. A let one critical comment slip out—which his husband caught—but Mr. A quickly shut down. Ben became frustrated by Mr. A's refusal to speak more about it; Ben then emotionally withdrew for a few days. As we spoke about this interaction, Mr. A said he was trying to be "benevolent." He felt rather guilty for criticizing his husband. He commented about his aversion to his husband's anger, saying that he was trying to behave toward his husband the way he wished his husband would be toward him. At some point, I interrupted him and said: *A bit ago, you said "I'm benevolent." What if we just took the first part of that: "I'm Ben"? I wonder if there's a way in which you're Ben?*

This intervention is an example of punctuation (Fink, 2011). Within Lacanian theory, speech is a chain of signifiers: one signifier follows another. Meaning is constructed retroactively, insofar as the addition of subsequent signifiers changes the meaning of the signifiers before. Meaning—the signified—forever slips underneath signifiers which themselves have no inherent connection to any particular signification.

Punctuation cuts the signifying chain at a certain point to highlight the polysemy of a particular signifier. It aims to open new possible meanings in *unexpected* directions, to shift from the self-reflective, egoic narrative that the patient is telling themself. A patient once described the deadlocks of the ego narrative wonderfully as "analysis paralysis." One attempts to think one's way through a situation and analyze all possible solutions but nevertheless ends up stuck. Some patients consult therapists because their current thinking through situations either leads to indecision or a decision that is not quite working. Punctuation aims to open new potential directions in a non-directive fashion (Israely, 2018).

Mr. A heard the polysemy in the punctuation *I'm Ben* and asked "Do you mean friend Ben or husband Ben?" I replied "either!" That session, Mr. A primarily spoke about ways in which he was *not* like husband Ben until ultimately, through the thickets of associations and reasonings, he reflected on how he was, in some ways, like his husband. Both shared a tendency to emotionally wall-off, albeit in different ways. The following session, we returned to the phrase *I'm Ben* and took up the other track, namely how Mr. A tried to be the rational advice-giver to his husband and to himself by "separating" his irrational, childhood feelings.

Two points of identification—his husband and the friend—were unveiled through the signifier *Ben*. To be precise, the punctuation *I'm Ben* cut the signifying chain *I'm benevolent*. The contiguous ordering of signifiers was adjusted, allowing new signification. The ego narrative (Mr. A painting his imaginary self-image as benevolent) was shaken up.

Another example of punctuation comes from relational psychoanalyst Philip Bromberg—although he does not use the term punctuation. This excerpt is discussed in the context of standing in an uncertain, "potential space" of not-knowing in analysis. Kate (Bromberg's patient) returned from vacation and reflected on how she felt much freer there than in New York. She wondered if this was because she felt required to tell Bromberg about everything and was always concerned with his reaction. She wondered whether this was because he liked her, and whether she was afraid to experience (or want) being liked. She drew a comparison to her surprising lack of chocolate allergy (and guilt) while on vacation, saying:

> maybe the truth is that you [Bromberg] are like chocolate to me. No matter what you say about me I can't take it in without getting pimples, because when I start to realize how attached I am to you, the pimples remind me not to trust you too much—to be careful of how much of me I show you— you could suddenly hurt me if I'm not who you expect me to be. (Bromberg, 1996, p. 528)

Bromberg responded that Kate was trusting enough to share this ambivalence of how much to share. Kate replied:

> I think you're saying that…because you are trying to get me to trust you more than I do. But I don't know if what I'm feeling right now is trust, or just a new feeling of 'I don't care what you think.' Right now, I really don't trust why you just said what you said. If I trust you instead of trusting *me*, I get pimples, and *that's zit*. (Bromberg, 1996, p. 528, emphases in the original)

Upon hearing the polysemic pun in Kate's speech—*that's zit* and *that's it* are near homophones—Bromberg burst into laughter. Kate, unaware of any intention to make a pun, joined in laughter, entering the "spirit of play" (Bromberg, 1996, p. 528). While Bromberg recognizes that his own bias toward slips may have led him to hear "*that's zit*" instead of "*that's it*," he nevertheless highlighted the ambiguity in speech via laughter. Bromberg interprets this moment as opening a playful "potential space" where previously disavowed issues could be explored.

Here, Bromberg punctuated Kate's speech through laughter. *Laughter was the interpretation.* His interpretation of this interaction as joining in a playful, freer, and uncertain potential space opens an important lens on Lacanian punctuation. Punctuating the signifier does not only serve to highlight key words in the patient's speech; it is also a practice of creative exploration of uncertainty. This is why Bromberg's laughter was an excellent punctuation. There was no meaningful (semantic) intervention—whereas my *I'm Ben* punctuation could be said to *imply* my telling Mr. A that I think he's like Ben. Playfully highlighting ambiguity opens the *potential* of what could be different, shaking up the narrative, and modulating self-reflective fixity. As an interpretation, punctuation *opens up* a novel space in speech.

Scansion

Scansion refers to the (in)famous Lacanian variable-length session. This does not simply mean a "short session." Scansion is the sudden ending of a session after an indeterminate amount of time, typically after something significant has been said. It is an analytic act that makes a "cut" into the patient's speech (Guéguen, 2012). It allows the analyst to make a more radical intervention that highlights a phrase and does not allow the patient to cover their tracks with further reasoning or rationalization. In some instances, scansion can function like a more dramatic punctuation and further highlight polysemic speech (Fink, 2011). In other moments, it can be used to fix or "quilt" a novel signification that has emerged (Israely, 2018). In both cases, scansion intends to keep the (un)conscious working in-between sessions.

For example, a patient described their experience with a new medication: "I feel better but also worse." I asked her to say more. She reflected that she felt "less afraid" but was more apathetic, commenting that "fear is a great motivator." I punctuated this phrase with curiosity and intrigue, and we discussed it in greater detail.

A few sessions later, she reflected on early childhood experiences with her mother, who was chronically anxious and "afraid," but also pushed her children to be "fearless." One time, during what was "supposed to be an encouraging speech" her mother's own anxiety got the better of her and she grew withdrawn and quiet. I commented: "A few weeks ago, you said 'Fear is a great motivator.'" She remembered the phrase and pondered it, wondering aloud "maybe that's not true." She reflected that it did not seem true for her mother. I suggested we end the session there.

In this moment, the phrase "Fear is a great motivator" took on a new signification from within the patient's own speech and life story. Ending the session aimed to quilt this novel meaning: namely, that fear might not be a great motivator. Like punctuation, it aims to shake up the patient's self-reflective narrative—highlighting not understanding but holes and inconsistencies.

Scansion hinges on the use of surprise to drive further (unconscious) work and change. One patient presented with frustrations that his family and friends never allowed him to have any independence. He was born with a physical disability but had adjusted reasonably well. Nevertheless, his parents—especially his father—insisted on taking extensive care of him. A few sessions into our work, he was speaking about one such instant and said, in a dejected tone, "I don't really know what else I can do. I'm not sure how to talk to them about this. It's not as if I can tell them to help me less, so…" I immediately interrupted him and said emphatically: "There! What you just said! That's a really good place to stop. I'll see you next week." He was a bit shocked, as this was the first sudden scansion in our work.

The next session, after discussing some similar themes, I asked him if he remembered where we stopped last session. He said: "Yes. I was saying how I wished I could tell people to not help me." I nodded enthusiastically and asked him what it was like for us to end like that. He said: "It surprised me" and that it was "not normal for me" to say something like that. This example illustrates scansion as a dramatic punctuation. Moreover, this scansion had the surprising effect (to me as well) of transforming a prohibition ("it's not like I can tell people to not help me") into a wish ("I wish I could tell people to not help me").

In her account of her own analysis with Lacan, Betty Milan writes on Lacan's use of scansion:

> Lacan made extensive use of the so-called short session. In fact, the time of the session was variable. Lacan was not guided by the time of Kronos, the time that passes, but that of Kairos, the moment of opportunity that we seize. What counted for him was the discourse of the analysand, not the clock…Lacan did not interpret the analysand's discourse by attributing a meaning to it. He interrupted the session and allowed the analysand to interpret the reason for the cut. The analysis continued outside the session. This new way of working was due to a clinical discovery. Lacan's method of working was based on the idea that the traditional way of interpreting provoked resistance. (Milan, 2024, p. ix)

The "traditional way of interpreting" could be understood as anything ranging from suggesting an explanation to an object relational commentary on undergirding fantasy life. Such interpretations—explanations—can be agreed with or disagreed with by patients at a self-reflective level: "Yes, I see that connection" or "No, I don't think that's quite it." It sits in a register of understanding (albeit perhaps a strange understanding). While interpretation may at times take this shape, Lacanian approaches emphasize another dimension of interpretation as provoking uncertainty rather than explanation. As Lacan put it: "Analytic interpretation is not made in order for it to be understood, it is designed to make waves" (Lacan, 1976a, p. 35). These waves are tides of potential and surprise at the level of speech.

A Note on Technique

Before I move on to considering scansion and punctuation neuropsychoanalytically, I wish to make a general comment regarding Lacanian technique. To be clear, I am not privileging punctuation and scansion over other types of interpretations. Such interventions do not work with all patients. Just as each case must be considered one-by-one, so too must technique (Guéguen, 2012; Milan, 2024; Miller, 2023).

It may be more helpful to consider the *purpose* of scansion and punctuation—rather than becoming caught up in the technique itself (although I do believe the technique is especially but not uniquely suited to this goal). The purpose is to *make waves* rather than meaning. This is best—perhaps only—possible through surprise and uncertainty. In a sense, this means that one does not know whether an intervention was an interpretation until after the fact (Busiol, 2021). We cannot be *a priori* sure what intervention will be taken as a surprise by a patient. Punctuations and scansions can still be replied to with "Oh, I suppose I can see that connection"—indicating that the cut was taken as a suggestion. Sometimes a question or something the analyst does not intend as an interpretation can (surprisingly) function as an interpretation, especially when the speech that follows are not what either patient or analyst expected (Busiol, 2021).

Other times, punctuation can become an intellectual engagement, where both patient and analyst become lost in the endless polysemies of language. Guéguen writes on this issue in some Lacanian circles:

> The passion for the formations of the unconscious [dreams, slips, polysemy, etc.] and the identification of the concepts of the unconscious to the laws of language (metaphor and metonymy) led some of Lacan's students to an erratic use of homophony, transforming in the worst cases the psychoanalytical treatment into an exchange of formations of the unconscious. (Guéguen, 2012, p. 12)

Interpretation can take many forms—from commentary on unconscious phantasy, to punctuation, to construction, to questions, to wondering about connections, and so on. While use of homophony might be a privileged mode of interpretation due to its potential to make waves and rouse enigma, Lacan comments that "the analyst makes use of it where suitable" (Lacan, quoted in Guéguen, 2012, p. 13).

The patient comes before the technique. For some patients, scansion may provoke too much anxiety or confusion to helpfully put the unconscious to work. As Milan reiterates throughout her memoir, one of Lacan's clinical motifs was to "*avoid the rupture*" (Milan, 2024, p. 19, emphasis in the original). Technique should not be adhered to at the expense of an irreparable rupture in the treatment.

With the recognition that punctuation and scansion are not the only ways to harness surprise, let us now consider these two Lacanian techniques neuropsychoanalytically.

Lacanian Technique in a Neuropsychoanalytic Lens

I claim that these Lacanian techniques have the aim of provoking prediction error. Specifically, they harness surprise to "make waves" in the patient's speech, to rattle it and shake it up (Fink, 2011; Soler, 2015). From a neuropsychoanalytic perspective, speech is a series of predictions—a *predictive chain*. Each signifier is a prediction that aims to deal with uncertainty at multiple hierarchical levels, including semantic, perceptual, and emotional free energy.

Interventions that highlight particular signifiers have the effect of shaking up this predictive chain, effectively questioning the semantic (predictive) resolution to the story. Punctuation and scansion accomplish this by breaking up the chain or extracting a particular word or phrase (e.g., a negated statement, a slip, or a comment quickly rushed over). These "cuts" to the predictive chain shake up the semantic resolution and open less intended routes. Insofar as constraining of associations is necessary to produce meaning (Bazan et al., 2021), Lacanian interventions work in the reverse, to shake up pre-established understanding.

Computationally, Solms characterizes language as "artificial manipulation of precision" (Solms, 2021b, p. 233). Words—as abstractions—can be linked to bottom-up input (cf. symbolic-imaginary knotting; see Chap. 9). Doing so increases the precision of bottom-up signals that might otherwise not be prioritized by drawing them into conscious attention and awareness. In other words, speech—by abstracting lower-level predictions—allows their precision-weights to be adjusted when they may have otherwise remained monotonous (and thus unconscious). Importantly, because speech utilizes shared predictions, these precision-weights can be adjusted by *others* (like an analyst) rather than only through bottom-up learning.

In Lacanian neuropsychoanalytic terms, interventions targeting the signifier-prediction (through punctuation or scansion) increase the precision of predictions, bringing them into consciousness. Solms highlights this consciousness as the self-reflective space of working memory that can *problematize* these predictions. I further highlight the fact that this consciousness is not only the space of self-reflective working memory but, more fundamentally, the empty space of $ in the contradiction of affective hyperpriors. That is, the consciousness brought to the highlighted signifier-prediction is not principally egoic consciousness but affective consciousness as felt uncertainty. Through surprise, punctuation and scansion *provoke* this uncertain space, to bring out an uncertainty that does not make sense (which one can then problematize). This surprise can be characterized as provoking *jouissance*.

One can be more precise regarding the Lacanian attention to words as signifiers—that is, the signifier as a motoric prediction (see Chaps. 7, 9, and 13). On the one hand, it *reduces* the precision of the motor signifier with respect to semantic prediction error. That is, there is an *increase* of semantic prediction error, the shaking up of the ego narrative. On the other hand, the *precision of the word itself is increased*. The word as a motoric form is highlighted as important, tagged with surprise. Prediction error is provoked while the precision (salience) of the word itself is emphasized.

I suggest that this process finds a neuro-computational parallel in the anatomy of predictions and prediction errors (see Parr et al., 2022). Predictions are projections downward to explain ascending prediction errors. Importantly, the motoric signifier as a constellation of neural activity can function as both. As a prediction, the motor signifier is tied to other non-declarative and emotional prediction errors. As a prediction error, the motor signifier must itself be understood semantically. Highlighting the motoric signifier as a prediction gives weight to the motor form as itself dealing with prediction error while loosening its precision as a (semantic) prediction error.

This allows me to differentiate reducing prediction error and *metabolizing* prediction error. Prediction—at least in the Fristonian computational definition—reduces prediction error. For Solms (2021b), increasing uncertainty is "bad" from an evolutionary-biological point of view. Correspondingly, reducing uncertainty is evolutionarily "good." The existential necessity to reduce free energy resonates with Holmes and Nolte's point that "the brain abhors informational surprise" (Holmes & Nolte, 2019, p. 1).

Psychoanalysis, however, particularly with its attention to the drive, complicates this straightforward perspective on uncertainty and prediction error. In their integration of the free energy principle with psychoanalytic psychotherapy, Holmes and Nolte reflect:

> there is an affective arc of motivated tension, consummation, and resolution, in which the very binding of energy is rewarding. By deepening trust and discouraging premature closure of surprise, therapy fosters this expansion of the realm of desire. Put another way–ambiguity and its resolution is both *exciting* and *rewarding*. (Holmes & Nolte, 2019, p. 9, emphases in the original)

Unsurprisingly, Holmes and Nolte come to this point when discussing sexuality. Sexuality is the space of *jouissance*, the "exciting" enjoyment in the increase of tension. By "discouraging premature closure of surprise," psychoanalysis opens "desire" and "ambiguity." Likewise, Rabeyron characterizes psychoanalytic therapy as "an interplay between extension and reduction of free energy" (Rabeyron, 2022, p. 1).

Drive—itself a complex montage (Johnston, 2005; Lacan, 1964) and situated in a sphere of conflicting affective hyperpriors (see Chap. 8)—complicates interpreting predictive work as the reduction of tension. I suggest that predictive work can also involve the *metabolization* of prediction error. Metabolizing prediction error does not eliminate it (as suggested by "explanation" and "resolution"). Rather, it sustains it and puts it to use, pushing the patient to do something with it. This is the essence of motoric prediction as action rather than understanding. A prediction that metabolizes prediction error *sustains* uncertainty, ambiguity, and desire. The metabolization of prediction error taps into the drive's enjoyment in the increase of tension.

Metabolizing prediction error thereby involves surprise—a certain ongoing engagement with uncertainty, a looseness of prediction. Lacanian techniques perform artificial precision modulation at an affective level, provoking affective attention and arousal through highlighting key signifiers. At least three precision adjustments might thereby occur: reduction of semantic precision, an increase of motoric precision (confidence in the motor form itself as important in the motoric predictive chain), and an assignment of SEEKING-precision via surprise. The surprising intervention—most dramatically with the sudden ending of the session—creates a space of novelty, SEEKING's bread and butter proactive engagement with uncertainty.

This is what happened in my scansion of "It's not as if I can tell them to help me less…" The predictive chain was interrupted, with an increase of semantic uncertainty. However, I highlighted the particular motor forms—"What you just said!"—as themselves important, more important than whatever words were to follow.[1] Likewise, the intervention had the effect of surprise (both computationally and in the patient's affective experience). SEEKING increased: his own curiosity of what he just said, how it was "not normal" for him, and led him to re-crystallize a new semantic understanding in the form of desire: "I wish I could tell them to not help me."

Such interventions motivate the psyche to perform *predictive work* to not only reduce but also *metabolize* prediction error. Broadly speaking, reduction of prediction error is a corollary to a self-organizing system in which this reduction is straightforward and piecemeal. However, when total reduction of prediction error is *impossible* (cf. affective consciousness as an effect of homeostatic antagonism), the corollary is not just to have "good enough" reduction. If J cannot be fully eliminated, one must metabolize this prediction error for *productive* and *creative* purposes.

---

[1] To be clear, I am not suggesting that those words *in themselves* were more important than what was to follow. Who can be in a position to judge which words are most important? The scansion performed an artificial precision modulation; the *intervention* rendered them more important.

Lacanian Intervention through the Lens of PLAY: Metabolizing and Modulating

I suggest that PLAY offers a helpful neuropsychoanalytic view on these Lacanian interventions. Solms reflects that "psychoanalytic treatment" is "at bottom, a form of PLAY (think of the balanced hierarchy, the reciprocity, the boundaries, the as-if quality of the transference, etc.)" (Solms, 2021a, p. 576). Kellman and Radwan (2022) review the neurobiological correlates of various forms of play, especially its intersection with the SEEKING system in the basal ganglia. They note that play "may have a central role in optimizing flexibility and creativity in individual response to novelty" through balancing different neural networks with (often) opposing responses or tendencies (Kellman & Radwan, 2022, p. 884). This fits with Solms's (2021a) conception of PLAY as a way of dealing with the conflicting systems involved in the Oedipus complex.

This PLAYful view resonates with Bromberg's potential space, particularly his laughter-interpretation, and the Lacanian use of word-*play* more generally. Highlighting the ambiguity of the signifier does not explain away the prediction error; it provokes the uncertain space of prediction error via surprise. PLAY allows one to "celebrate the joy of surprise" (Kellman & Radwan, 2019)—an *enjoyment* in the ongoing space of prediction error.

This perspective can link to Lacan's own reflections on humor:

> But if the truth of the subject, even when he is in the position of master, does not reside in himself, but, as analysis shows, in an object that is, of its nature, concealed, to bring this object [*objet a*] out into the light of day is really and truly the essence of comedy…humour…is simply the recognition of the comic. (Lacan, 1964, p. 5)

The "truth of the subject" resides not in predictive mastery (the domain of the ego, some internal knowledge one can master) but in an *objet* that is typically concealed by imaginary gymnastics and reasoning. Sometimes, the unveiling of the *objet* causes anguish (Soler, 2015), but here Lacan designates another modality of its exposure: humor.

Recall that *objet a* stands for the surplus within the predictive field, its insistent remainder of prediction error. Insofar as patients complain of consciously felt affects, they are pointing to an excess that they cannot bear despite their current predictive resolutions. There is a challenge in how they are situated toward *objet a*. Here we (re)find the formula for the fundamental fantasy: $\$ \lozenge a$. The subject *feels* their division in the face of *felt uncertainty*.

Comedy—as a social form of PLAY—allows one to detoxify or *modulate* (see below) the stance toward *objet a*. The surplus remains, but one can tolerate it, put it to productive use, and—in the best of cases—enjoy the enjoyment (Fink, 2011; Israely, 2018). I will return to this point below regarding formulation. As Bromberg's intervention illustrates, this is why humor is often an exceptional intervention. It is a PLAYful space in which patients can question and problematize the predictions they hold so dear. To paraphrase Mark Epstein (2019) on Buddhism and psychoanalysis: 'The idea of enlightenment as the disappearance of the ego does not mean that there is no self. It means that your ego, your sense of self, your sense of who you are, is a bit of a joke.'

PLAY, as a way of opening flexibility in how one responds to novelty (Kellman & Radwan, 2022), involves more than metabolizing prediction error (the "joy of surprise"). It also involves the *modulation* of predictions. In my view, metabolizing prediction error is the corollary to homeostatic impossibility and affective consciousness as irremovable J. Modulating predictions would then be the corollary to the impossibility of removing non-declarative (repressed) premature automatized predictions. They are indelible, firmly consolidated action plans; the repressed is doomed to repeat. However, I propose that the Lacanian wager is that *how* the repressed repeats *can* change—in line with the PLAYful shift in *how* one responds to novelty. The modulated prediction metabolizes prediction error *otherwise*. I will return to this point in the final section.

Neuropsychoanalytic Contributions to Lacanian Technique

In line with the dialogic structure of neuropsychoanalysis, here I suggest some contributions to Lacanian technique from this intersection with neuroscience. At the outset, I wish to characterize these suggestions as just that: suggestions. Moreover, these ideas could be derived from thinking that does not rely on neuroscience. However, neuroscience allows us to consider them in a new, potential space.

Some Lacanian analysts (e.g., Guéguen, 2012) point out that Lacanian interventions are not only about word-play, polysemy, puns, and so on. This is sometimes how Lacanian technique is characterized (e.g., Busiol, 2021; Fink, 2011) in addition to the traditional Freudian emphasis on dreams, slips, and so on. One can characterize these interpretations as symbolic, insofar as they focus on the signifier.

With my Lacanian neuropsychoanalytic twist, there are *levels* of the symbolic (see Chap. 13). Symbolic interpretations rely on symbolic structure: a differential system organized in a contiguous chain that produces effects through metaphoric quilting. If the symbolic structure can exist with *any* differential system (not just language), then could one not apply the logic of symbolic interventions to other differential systems in the brain's predictive hierarchy? I have in mind here gestures (the motor system broadly speaking) and affects.

Repeated signifiers are often punctuated and highlighted because those repeating signifiers may hint at how the patient metabolizes *jouissance*. Can one not do this with affects that repeat? Consider the following excerpt from Melanie Klein. A patient tells her

> That he slept alone last night and wasn't the noise of the rain awful. Then gives me the reasons for the quarrel with his wife, who was going away for the weekend and he felt deserted by her. I interpret his despair about the good mother who left him, deserted him, and that he was taking this as a punishment…[H]e continuously says that it is quite clear that if he dreams of a person coming out of the grave it must be his mother, and that this very kind man stood for her. I suggested his feeling of despair, of being left alone by his wife, and by me over the weekend, and also of the holidays

coming near, connects with feeling of loneliness every night, when he was afraid of his mother's death, and felt left alone. (Melanie Klein, quoted in English, 2023, p. 36)

Among other things, Klein highlights the repetition not of a signifier but of a feeling: "deserted," "despair," and "loneliness." This points to the pattern of *how* this patient prioritizes conditions of uncertainty—that is, which emotional system tends to be prioritized and the contexts for its prioritization. One can attend to affects operating like signifiers, insofar as they too can be displaced, condensed, substituted, repeated, and so on.

Here, the notion of J(E) becomes helpful. I am not suggesting a straightforward belief in the consciously felt emotion because this affect can be displaced, condensed, substituted, and so on (see Chap. 12). This risks reifying the specific emotion (the self-reflective, imaginary level) rather than questioning *why* this particular emotion has been prioritized (the symbolic, structural level). The significance of the consciously felt emotion is the fact that *that* category of uncertainty is how the patient is currently prioritizing the surplus prediction error of affective consciousness. There is no pure J, but J as the disruptive surplus within the prioritized emotional system: J(E).

I suggest tracking the consciously felt *chain of affects* as a way of tracking *how jouissance* tends to be metabolized and prioritized. One could write this as: $J(E_1)$-$J(E_2)$-$J(E_3)$-... It is attention less to the particular emotion and more attention *to the fact of affective consciousness itself*. It fits with the question: *why* is *this system* prioritized, and could it be prioritized differently?

Recall Mr. B who presented with monolithic PANIC. Carefully, we unpacked affective sequences: sometimes J(LUST) then J(PANIC); other times J(RAGE) then J(PANIC); guilt at certain points [J(RAGE+PANIC)]; a noticeable substitution of J(PANIC) for what was once J(PLAY). For Mr. B, the predominant affect was indeed PANIC. Yet, the clinical work revealed a *chain* of affective systems with *patterns* of prioritization. The above sequences of prioritization were *feeling patterns* that repeated (analogous to signifiers that repeat); other patterns such as J(PLAY) then J(CARE) were uncommon or absent.

One can ask whether an *affective punctuation* might be one angle of intervention. With Mr. B, we took space to pause and highlight one affect before its quick displacement onto PANIC. This method is already formalized in mentalization-based therapies (Fonagy et al., 2002). This could interrupt the *affective chain* and produce new possibilities of prioritization. Notably, this involves more than a different feeling; it involves the adjustment of precision-weights throughout the hierarchy in accordance with the newly prioritized emotional system.

Additionally, one can chart repeated gestures, treating them like repeating signifiers. A patient began a session by talking about how sexually frustrated they were with their partner. As they spoke, they made a dramatic, downward, chopping gesture with their hand to emphasize how they sometimes did not want to have sex. Vice versa, their partner was never in the mood when they were. Later on, I asked if they had any dreams or daydreams. They shared how they were fantasizing about doing martial arts while at work. They had not practiced for several years but were very involved in childhood. I asked more about the martial arts fantasy, and they demonstrated some techniques with their hands. One gesture was a similar chopping motion.

I pointed out that they had made the same gesture in relation to their frustration with their partner. He was intrigued, and we wondered how the combination of physical (and mental) discipline in martial arts with intense physical expression contrasted with the mismatch of sexual arousal and when one was "supposed" to want to have sex.

This example illustrates the punctuation of a repeated gesture (rather than a word). It had a similar effect of stirring curiosity and novel considerations. Generally, one might wonder why a certain gesture was repeated for two stories.

These clinical suggestions follow from the threefold movement method (see Chap. 3). One maps the signifier in neural space. There, we find that the logic of symbolic organization applies not only to words but to affects and gestures as well. From there, we return to psychoanalytic space to consider whether interventions aimed at the signifier can be adapted to general motor gestures and affects. I hope that this approach exemplifies how neuropsychoanalytic dialogue opens up clinical possibilities and considerations. It should be clear how dialogue with neuroscience in no way ushers in a normative clinical stance.

Lacanian Neuropsychoanalytic Clinical Formulation

Here, I draw connections between Lacanian clinical formulation (especially Miller, 2023) and Solms's (2018) model. I additionally put forth a Lacanian neuropsychoanalytic version of clinical formulation that integrates the various concepts from this book.

Charting J among the Constellation of Affects

I have already discussed how a Lacanian neuropsychoanalysis prioritizes the logic of *jouissance* in particular emotional systems. This refers to the idea that the consciously felt emotion is not simply an unmet need; it is itself enjoyed, operating not homeostatically but in the logic of repetition of this enjoyed tension. Obstacles to homeostasis are not only hard to change because they are deeply automatized. They resist change because they are enjoyed—they are how the subject metabolizes *jouissance*.

Where Solms (2018) recommends identifying the predominant conscious feeling, a Lacanian neuropsychoanalysis would privilege an ongoing charting of *jouissance*. By this, I mean discerning the shifting vicissitudes and constellations of affective contradiction. Affective hyperpriors *will* conflict; *how* they conflict—how they tend to be prioritized—will vary in each case.

In some cases, there may indeed be a prevailing affect. In such situations, charting J would ask why this particular affect's prioritization is so generalized, which affects might be avoided through its generalization, and so on. Or, consider Klein's charting of conflicts between RAGE, PANIC, CARE, FEAR, and so on—to insert neuropsychoanalytic terms in to Kleinian psychoanalysis (Dauphin, 2023). Specifying how these systems play out in an individual's case and history reveals *how* the contradictions of affective consciousness play out. Is there especially a conflict between RAGE and FEAR? Or between SEEKING and PANIC? Is a conflict between LUST and FEAR "resolved" by prioritizing RAGE? These details are significant because they structure predictive echoes in the transference, symptoms, interpersonal repetitions, and so on. One can do this by following the *jouissance*, the surplus in each emotional system—J(E)—specifically how the systems are *aberrated* and follow the logic of excess. Which systems (or series of systems) tend to repeat?

While this differs from Solms's recommendation to ask "what is the conscious affect?", I do not believe it departs very far from Solmsian formulation. When discussing cases, Solms carefully charts the vicissitudes of different systems (e.g., Smith & Solms, 2018; see below). He also recognizes that multiple systems will be dysregulated in every case (Mosri, 2023). What my perspective adds is the emphasis on *jouissance* as an effect of contradiction, attending less to the particular quality and more to the conflictual landscape in which *jouissance* is situated.

Symbolic Reduction: S1 and the Fundamental Fantasy

With our Lacanian neuropsychoanalytic mathemes, we can be more precise in this landscape of *jouissance*. I have suggested that Solmsian premature automatization corresponds to the Lacanian symptom and can be written as S1-J, the knotting of a master signifier to *jouissance* emergent in the sphere of conflicting affective hyperpriors (see Chap. 9). I use the matheme S1-J rather than S1-J(E) to specify that the prematurely automatized action plan (S1) is situated in relation to the *constellation* of conflicting emotions (J), not just the most salient or prioritized emotion [J(E)].

Consider the following clinical example (for full details and discussion, see Smith & Solms, 2018). A patient presented with severe depression, guilt, a history of relationship troubles (cheating, being cheated on, etc.), and a history of sexual abuse at the hands of his foster father. Commenting on the childhood details, Solms notes that this patient did not flee from his abusive foster father: "The affect that should accompany this instinctual problem (i.e. the need to remain safe from predators and other such dangers) is fear…why did he not feel fear, first and foremost: why was this not his presenting complaint?" (Smith & Solms, 2018, p. 57).

For Solms, the fact that the presenting (i.e., conscious) affect is depression points not to FEAR but instead to attachment (PANIC's grief-phase, downregulating SEEKING): "Depression...is something one feels when one loses hope about the prospect of achieving reunion with a love object" (Smith & Solms, 2018, p. 57). Given that depression is consciously felt, this is where Solms situates the prematurely automatized prediction. With recognition of several childhood scenes between this patient and his foster mother (especially one where she threatened to throw him out of the house after a peeing accident), Solms infers:

> His deeper, repressed prediction—automatized as early as 3 years—seems to have been something like: "if I am really good, I will not be abandoned"; "if I never protest, mummy will not reject me." So his repressed action plan, his solution to the problem of separation distress, seems to have been simply this: *do nothing*. That is not a very good prediction; it is bound to end in tears...That is why our patient suffers from depression; his solution to the problem of separation distress (to the instinctual need for reunion) did not hit the mark. (Smith & Solms, 2018, p. 57, emphasis in the original)

Do nothing is the prematurely automatized solution to PANIC. However, Solms relates this automatized prediction to more than just PANIC. Rather than flee in FEAR from the abuser, this patient *did nothing* because the abuse also evoked PANIC:

> he fell back upon his preexisting (repressed) solution*: he did nothing*. Since this, too, inevitably, didn't work (cf. prediction error), he was left at the mercy of the affects of PANIC *and* FEAR. (Smith & Solms, 2018, p. 57, emphases in the original)

Solms conceptualizes this patient's secondary defensive prediction to deal with FEAR and PANIC as "turning PANIC into LUST," himself becoming a seducer (Smith & Solms, 2018, p. 57). However, this led to directing RAGE against himself that should have been directed toward his abusive foster father. Hence, he also presented with guilt.

Solms's formulation wonderfully illustrates how the prematurely automatized prediction is situated in relation to a *constellation* of conflicting affects (here, at least PANIC, RAGE, LUST, and FEAR). *Do nothing* is indeed a poor solution for PANIC. However, *do nothing* is a poor solution for *all* of Panksepp's systems and is likely to fail in *many* human situations. As an S1, *do nothing* repeated not only for the PANIC situation (with his foster mother) but also for the FEAR situation (with the abusive foster father). In a disguised fashion, it persisted through the secondary defense of becoming a seducer. Moreover, *do nothing* had the signification of "do not protest" and "be a good boy," which was how it repeated in the transference. *Do nothing* is an S1 charged with *jouissance*. It is knotted to surplus prediction error and is how this patient metabolizes (not reduces) *jouissance* as prioritized in multiple conflicting emotional systems: S1-J.

Solms (2018) calls for a continued pointing out and problematizing (the derivatives of) the repressed prediction. First, note that this differs from calling for an expansion of possible meanings or the meaning-making process of analysis. Rather, this technique aims at isolating a repeating pattern.

I suggest that this corresponds to "symbolic reduction" in Lacanian terms (Lacan, 1956–1957, p. 166). For Lacan, (symbolic) interpretation ultimately "has the effect of bringing out an irreducible signifier" (Lacan, 1964, p. 250). Rather than (solely) proliferating the endless possibilities of meaning (which would be an imaginary expansion of signification), symbolic interpretation "is not open to all meanings" (Lacan, 1964, p. 250). Through opening the polysemy of speech, novel signifiers and significations emerge, but so too does a pattern. Certain key signifiers *repeat*. Free association leads to repetition. In symbolic interpretation, the subject "should see, beyond this signification [of meaning]…to what signifier—to what irreducible, traumatic, non-meaning—he is, as a subject, subjected (Lacan, 1964, p. 251).

Miller characterizes this mode of Lacanian analysis as "a *reduction-operation*," which he opposes to "*signifying amplification*" (Miller, 2023, p. 23, emphases in the original). Symbolic reduction does not (endlessly) follow the structure of proliferating new meaning (although this is a necessary dimension; simply put, one must allow amplification for the crucial pattern to emerge). It has the structure of condensation in a joke or *Witz*. Following "repetition leads to a reduction operation that is a formalization," insofar as repetition leads to a convergence on a certain axiom that insists irreducibly (Miller, 2023, p. 30).

Formulaically, one can call this the repetition of S1s. Proliferated association to many scenes, fantasies, dreams, symptoms, and so on converges (is reduced) *through* the repetitions. Punctuation and scansion that interrupt the chain orient toward a symbolic reduction beyond signification. Miller notes: "the reduction-operation is therefore the statement of convergence which is a *master-signifier* [S1]—the signifier that has become master of the subject's destiny" (Miller, 2023, p. 33, emphasis in the original).

A Lacanian neuropsychoanalytic view helps clarify Lacan's characterization of this S1 as "beyond…signification," an "irreducible signifier" to which the subject is "subjected." Beyond signification points to the beyond of meaning—that is, beyond declarative (semantic) predictions. It aims at the non-declarative level of motoric predictions. Moreover, it is something to which the subject is *subjected*—doomed to suffer its repetition.

Symbolic reduction aims (in part) at the isolation of the premature automatized prediction as an S1. Insofar as symbolic reduction identifies premature automatization as an irreducible prediction, it homes in on the *mode of enjoyment*—the subject's means of deriving *jouissance*. It is *how* the subject deals with the problem of the drive and the contradictions of affective consciousness.

Discerning S1-J in a case can be formulated as considering how the prematurely automatized prediction (S1) fits into the constellation of contradicting emotional systems (J). Symbolic reduction to S1-J attempts to situate *how* S1 *does not work* for *multiple* affects yet paradoxically demarcates their repetition and metabolization of *jouissance*. In the case above, *do nothing* is not a very good prediction, but it is nevertheless the motoric prediction tagged to deal with surplus prediction error (affective consciousness).

Recall the case of Mr. A. Let us suppose that "separate" is Mr. A's repressed prediction. It originated in the advice from a man on how to deal with the FEAR of the snake but repeats in derivative scenes such as his relationship with his husband, his work, and so on. For example, his inhibition of his RAGE due to FEAR of his husband's RAGE sits alongside CARE for his husband and attachment needs (PANIC). However, his actions follow a plan of *separating* his own emotional response. *Separating* fits into a particular constellation of emotional conflicts with excess uncertainty and repetition: S1-J.

This example (and others; e.g., Smith & Solms, 2018; Mor-Ofek, 2022) points to a certain nuance in the quality of repressed predictions. Solms (2018) characterizes the automatized repressed as a non-declarative prediction, analogous to procedural memories. However, the repressed prediction is not the same as standard procedural memories in neurobehavioral literature like "how to play a piano" or "how to ride a bike." They are more abstract (and relational) in nature.

Do nothing can take a variety of forms. One can literally do nothing, talk to someone else instead of a loved one, become hyper-involved in a hobby to the exclusion of social life, refuse to text, or perform all sorts of cognitive and interpersonal gymnastics to precisely prevent anything from happening. Repressed predictions repeat in their derivatives with much greater complexity and variety compared to procedural memories like how to play piano.

In other words, there is the repressed at one level, then there is the level of *how* the repressed repeats. I have already proposed that the fundamental fantasy (as an abstract hierarchical control rule) nests the prematurely automatized prediction and governs its context-specific implementation in derivative scenes (see Chap. 9). There are two levels: the direct knotting of S1-J in the symptom and the symbolic mediation of the fantasy (Miller, 2023).

With this Lacanian neuropsychoanalytic lens, one can attend to Solms's different formulations of the repressed prediction in the case discussed above. One is "if I never protest, mummy will not reject me"; another is "*do nothing*" (Smith & Solms, 2018, p. 58). Likewise, in the FEAR situation, "do not protest" becomes "do not flee him" (Smith & Solms, 2018, p. 58). The difference here is not simply Solms's diction; it is the difference between conditional logic (if…then) and a motoric form *per se*.

The fantasy (in its imaginary dimension) is the (childhood) scene: *if I never protest, mummy will not reject me*. The symptom—*do nothing*, as mode of enjoyment—is nested within the fantasy (indeed, Solms comments on masochism condensed here). In Lacanian fashion, the formulation *if I do nothing (or substitute for doing nothing), then the Other (or substitute figure) will not reject me (or substitute for rejection)* allows one to recognize the abstract schema through which the automatized repressed finds its derivative, declarative expressions.

This raises the question of whether, when inferring the repressed prediction, one is inferring the *symptom* or the *fantasy*. The fantasy enables the repressed prediction (S1) to have a different signification, so to speak. Hence, "do not protest" can become "do not flee," both indexing the S1 *do nothing*. Reversals can likewise take place, where S1 repeats in an inverted fashion, as excessive seduction, for example. Hence the utility of the Lacanian fantasy formula as $\$\lozenge a$. The $\$$ finds a relation to the uncertainty in the general predictive field (Other), the variety of emotional prioritizations [J(E)] that "bulge" (*objet a*) in different relationships. The subject *feels* this division, not in a single emotional system, but in the excess excitation from *conflict* between different systems whose prioritizations are constellated around a (repressed) repeating prediction (S1).

Differentiating fantasy as *mode of relation* from symptom as *mode of enjoyment* thereby has significant clinical implications. While the prematurely automatized repressed will repeat, it might be *modulated* such that *how* it repeats (through modulations of the fantasy's superstructure) can change (Dall'Aglio, 2023). Such modulations have the potential to *modulate* the symptom to a singular mode of enjoyment. The S1 still metabolizes prediction error, but—by modulating the prediction—it metabolizes it otherwise. One makes do with one's symptom. Let us unpack this point.

Real Modulation and Metabolization: From Symptom to *Sinthome*

Miller refers to the real as distinct from the symbolic to highlight this difference between fantasy and symptom:

> Beyond the symbolic reduction of the highlighting of compressed formulas and of emergences of the particular oracles for each person…Why has such and such a word from the Other—from the father, the mother, etc.—taken

on a decisive value for the subject? Why did a particular misunderstanding or homophony hit the nail on the head? (Miller, 2023, pp. 42–43)

Here, Miller asks *why* a particular S1 (as symptom or fantasy) takes root for a subject. Symbolic reduction to neither the fantasy nor symptom explains *why* it repeats. This is one step away from asking whether and how the repetition of the S1 can change. For Miller, it is an issue of *jouissance*, of libidinal investment at a point of contingency.

Translating into Lacanian neuropsychoanalytic terms, we can say that the idiosyncratic symptom (and fantasy) take root because it metabolizes the J of antagonistic affective hyperpriors. This is the point of contingency that results from homeostatic impossibility. Miller argues that the "question is not so much to isolate them [certain formulas], but to know how [the subject] comes to yield the jouissance they bring him" (Miller, 2023, p. 48).

"Yield the jouissance" is the key phrase here. "Yield" indicates the limit of what cannot be reduced (symbolically interpreted or, in our terms, minimizing free energy). Solms recognizes this limit as the irremovable risk of the regression, the re-emergence of the prematurely automatized prediction. With the Lacanian neuropsychoanalytic terms here, however, I suggest we can go a step further.

At this point, Miller recalls Freud's note that drive-pressure exerts a constant force. If *jouissance* is impossible to eliminate, then symbolic reduction must be followed by something else, the *yielding* of *jouissance*, a making do with one's mode of enjoyment. Here, the difference between fantasy and symptom demonstrates its clinical utility:

> the signifier does not have a mortifying effect on the body, which is what the theory of fantasy supposes. What is essential is that the signifier is a cause of jouissance. It does not therefore attract libido but produces it in the guise of *surplus enjoyment*…The signifier fundamentally has an effect of jouissance on the body. Lacan called it the symptom.
> In this sense, the symptom goes beyond the fantasy. The fantasy supposes the body mortified by the signifier, while the symptom refers to the body vivified by the signifier. (Miller, 2023, pp. 56–57)

Fantasy "mortifies" the body through the signifier. In our terms, fantasy as an abstract hierarchical control rule reduces *jouissance* (surplus prediction error) through an extensive predictive hierarchy. The subject is situated as lacking ($) in relation to the remainder in the predictive field (*a*).

On the other hand, the body "vivified by the signifier" indicates how a subject yields their mode of enjoyment. I interpret the subsequent step (attuning to symptom after symbolic reduction) as the *modulation* of symptom into *sinthome*. *Sinthome* is a Lacanian neologism for a symptom to which the subject has a different stance, the symptom as knotting real, symbolic, and imaginary in a new fashion (Lacan, 1975–1976). In our Lacanian neuropsychoanalytic terms, modulation would occur through a shift in the fantasy structure such that the control hierarchy allows the repressed prediction to repeat differently, to have different derivative significations. Such change would occur through *problematizing* predictions and *provoking* prediction error to drive reconsolidation of the fantasy. Analysis ends with a shift "from the symptom, source of suffering, to the *sinthome* source of creation" (Busiol, 2021, p. 29). It yields *jouissance*.

Adrian Johnston describes *sinthome* as follows:

> Lacan stipulates that a *sinthome* is a symptom upon which the very being of its subjective bearer depends. Were the subject to be "cured" of his/her *sinthome*, he/she would cease to exist…Hence, the therapeutic gain brought about by analysis…hinges not on eliminating the *sinthome*…but on making it transition from being an "in itself" to a "for itself" (to resort to a bit of Hegelese not foreign to Lacan). In so doing, the subject goes from being unconsciously in the grip of his/her *sinthome* to having a margin of conscious distance from it, after the achievement of which he/she may even come to identify with it (or at least be comfortable enough living with it). This might be as much self-transparent freedom and contentment as analysis can afford. (Johnston, 2019, pp. 181–182)

I suggest that one can understand the shift from symptom to *sinthome* as a shift from a rigid means of enjoyment—a certain repetitive form in which the prematurely automatized prediction repeats—to a new, more flexible, and more self-accepted (identified, recognized) mode of enjoyment.

Importantly, the identification with one's *sinthome* differs from the imaginary identifications of the ego. The latter involves guises of identity, firmly held predictions that provide certainty (and are, in this sense, mortifying and limiting). Identification with the *sinthome* sustains a level of uncertainty—the opaqueness of the subject's *jouissance*, which is simultaneously vivifying—and recognizes declarative predictive resolutions as something of a joke (cf. Epstein's comment about the ego as a bit of a joke). Concerning this trajectory in analysis, Lacan comments:

> The mirage of truth, from which only lies can be expected (that is what, in polite language, we call 'resistance'), has no other term than the satisfaction [enjoyment] that marks the end of the analysis. (Lacan, 1976b, p. viii)

The enjoyment in the *sinthome* recognizes a *mirage* of truth, truth as a half-saying insofar as no truth can say it all. No prediction can eliminate all uncertainty. *Sinthome* marks a certain enjoyment that can be put to use while recognizing the inevitable short-comings of predictive work.

I suggest that this modulation of symptom into *sinthome* occurs through PLAY. PLAY is a means of creating flexibility among multiple neural systems regarding how one responds to novelty (Kellman & Radwan, 2022). Novelty is a hallmark of *jouissance* as surplus prediction error (surprise). Humor—which depends on surprise—likewise depends on extensive networks involved in modulation of reward-predictive (i.e., SEEKING-oriented) links (Mobbs et al., 2003; Shibata et al., 2014). Moreover, PLAY can function as a super-ordinate system to resolve contradictions among conflicting emotional systems (Solms, 2021a). For Solms, PLAY is the means for learning and forming rules; in the aftermath of the Oedipus complex, these are particularly the rules for managing different emotional systems. Insofar as I connect these rules to the fundamental fantasy, these rules would also involve predictions of precision, the patterns in the prioritization function of the midbrain decision triangle, informed by abstract rule hierarchies (see Chap. 12).

PLAY—by harnessing surprise to drive reconsolidation—can establish *new rules*, new rule-hierarchies or a reduced fantasy-hierarchy that modulates the automatized prediction such that it repeats in a different fashion. By hitting upon the more direct link between signifier (S1, action plan) and *jouissance* (J, affective consciousness, surprise) through reduction of the fantasy, PLAYful interventions can open a potential space, a creative space at the level of premature automatization. This can allow the symptom to shift into *sinthome*, a site of vivification and creative production in relation to the inevitability of affective consciousness. PLAY modulates the automatized prediction through the creation of new rules, new symbolic mediations of *how* the repressed repeats. This would include changes in the predictions of precision housed in the fantasy structure. One can employ the same action plan, but the situation might be prioritized otherwise—say as SEEKING instead of PANIC. There might be a shift in the *constellation* of conflicting emotional systems to the degree that the subject finds a way to PLAY with how S1 is knotted to J.[2]

To be clear, *sinthome* as creative production within a new rule system is not done *for the Other* (Verhaeghe, 2019). Either consciously through pedagogical intervention or unconsciously through one's history, a shift in enjoyment still indexed to the dictates of the Other remains alienating. In our terms, such a modulation *cannot occur through predictable means*. In other words, such a shift *must* emerge through a radical confrontation with uncertainty throughout the predictive field (i.e., the Other as barred and lacking, not fully deterministic; cf. alienation and separation in Chap. 9).

How can one conceive of new rules emergent through *uncertainty* in the predictive field of the Other? Is the reliance on an albeit reduced or modulated fantasy not still a fantasy that invokes the Other? Žižek provides the answer through his commentary on human games (implying PLAY) and their rules:

> Game is something…out of which you can step, in some sense, break the rules…in this sense, games are on the one hand cliches…My favorite quote from the big Hollywood…Sam Goldwyn, producer…his assistant tells him that in the press they say "there are too many old cliches in your movies." You know…how he answered? He sent a memo to his screenwriters, scenario-writers: "We urgently need more new cliches!" The right answer…Language is a game. Not just in this general Wittgensteinian

---

[2] I owe the origins of the idea of playing with automatized predictions to a conversation with Elizabeth Winship.

sense but in a more specific sense. How is something truly new created? When you play the standard language game, you say something wrong, you made a mistake, and then, instead of apologizing, you quickly invent a new sense to accommodate that mistake. So I think that all true…progress comes through this…A French friend of mine claimed to me, you see this is cooking, kitchen as game, why is French cuisine one of the great ones? Because it's all based on failure…Your ancestors, you were doing cheese, the cheese got rotten, you were lazy, but you say "My God, a new kind of cheese! Roquefort, or what." And you were doing wine…whatever happened to him…"Oh my God, let's call this champagne!"…I think this is the most productive way. The true creativity is not "Oh, I invented up something new" but precisely this, to quickly change…the rules so that an apparent mistake allows you to open up a new higher level of gaming. (Žižek et al., 2023)

The creativity at stake in the PLAYful modulation of the repressed and the formation of new rules that allow different repetition is not a top-down, educative learning or thinking process. The creativity of the *sinthome* follows the logic of a mistake, the potential space of spontaneity, surprise, prediction error, uncertainty, and PLAY. Hence the orientation of Lacanian techniques toward surprise and provoking prediction error. The "new higher level of gaming"—the new rules—are novel predictions that create a new predictive field where the repressed repeats differently. It is for this reason—*jouissance* in the space of contingency—that there can be no formula or universal technique for modulating the automatized prediction. One cannot predict which flexibly explored rules will become the "new cliches."

This is how one can understand the premature automatized prediction as a failed prediction in Solms's sense of failing to satisfy homeostatic needs. It will always repeat and thus continue to fail. Psychoanalytic treatment follows the logic of the famous quote from Samuel Beckett's *Worstward Ho* (often repeated by Žižek): *Try again. Fail again. Fail better.* The modulated automatized prediction still fails (i.e., continues to metabolize rather than remove *jouissance*), but it is a prediction that can then fail better (as *sinthome*).

A final example from Betty Milan's account of her analysis with Lacan can illustrate these concepts. (The interested reader is encouraged to read her moving account of her analysis.) For background, it suffices to say that Milan's analysis with Lacan dealt with a variety of issues, including anguish, depression, her relationship to language (Milan, raised in Brazil, spoke Portuguese but went to France and conducted her analysis with Lacan in French, which she had to learn), her family history (parents being of Lebanese descent; ingrained issues of racism and xenophobia), and her own desire to be an analyst in Brazil. Toward the end of her analysis, Milan recounts a session:

> "Your shirt reminds me of a sweet that I ate as a child, at the circus…and also of another one that exists only in my country." [says Milan]
> "What is it called?" [asks Lacan]
> "*Pé de moleque.*"
> "Mo-lek…?"
> "It means smart kid. A sweet that all Brazilians know and that no Frenchman can imagine. One more reason to go and live in my country…"
> "Of course."
> "You always agree, but you don't understand what I'm saying. I don't want to live here anymore, I'll go away. No word in the French language makes me dream. For me, French words are like objects: I keep bumping into them. Portuguese words are translucent, light as veils…it's the veil I want."
> "Well, my dear, see you tomorrow." (Milan, 2024, p. 61)

Lacan's "of course" here is ironic. He speaks as a Frenchman who does not understand Portuguese or Brazilian culture in depth. The ironic agreement provokes prediction error, a provocation that Milan takes up. This provocation, importantly, leads her to name her desire. However, this is a particular desire that—as Lacan likely intuited—has to do with a certain mode of enjoyment (Milan is a writer and translator as well; language is incredibly important and salient for her). Milan names her desire for words that are not opaque but "veils." At this point—the emergence of something novel—Lacan scanded the session.

This was the last session Milan had with Lacan as an analysand (there is a subsequent session as a supervisee). Following this session, Milan reflects:

The word *veil* made me dream about my father and the odalisque—harem concubine—costume that I had worn several times as a child for the carnival. When I woke up, I did not understand what the odalisque was doing in my dream. Did it remind me of the Sultan? The Eastern origins of my father? Or his insane jealousy? He always wanted to keep me to himself: it was as if he was forcing me to wear a veil. Suddenly, everything became clear. I finally understood why I had taken on an analyst whose language was not my own. I needed someone in front of whom I could not reveal myself completely...Despite myself, I had not ceased to be the object of the father's desire. Paradoxically, I had not chosen Lacan for what he knew, but for what he did not know, what he could not even be brought to know. (Milan, 2024, p. 63)

First, note that Lacan's scansion, on the emergence of the signifier "veil," led Milan to dream—contrary to her egoic narrative assertion that French words did not make her dream. Moreover, we can speculate that *veil* operates here as an S1 for illustrative purposes.

Let us suppose that "veil" (with significations such as predictive cascades to hide, cover up, perform, and so on) is a premature automatized prediction (S1) to the constellation of competing affects (J) in the family context, especially her father's desires and insecurities. Milan's dream is unconscious predictive work, following the scansion, that leads her to articulate a certain relationship to "veil" not as reducible to (alienation in) her father's desire to "keep me to himself." Rather, "veil" opens onto a new relationship not (only) to meaning but to language itself: Milan names her desire for the Portuguese language (as a tool of practice, of writing, of speaking, of analyzing) as a novel modulation of veil. Veil can take on a more flexible stance in the practice of writing—words as translucent, not opaque—rather than hiding. Lacan, as someone who could not understand her language, was crucial to this emergence. There is a modulation of the automatized prediction *veil* that allows its flexible emergence as something other. *Not understanding* (i.e., surprise, uncertainty, prediction error) in the analysis was crucial to this change.

Predictive Destitution

Some clarifying comments are crucial to not misunderstand my suggestion that the modulation of automatized predictions (as a means of shifting from symptom to *sinthome*) can occur through PLAY. As understood in the neuroscientific literature referenced above (Kellman & Radwan, 2022; Panksepp, 1998; Solms, 2021a), PLAY is an emotional category of flexibility, social joy, and interpersonal reciprocity. While these features *can* be aspects of the PLAY which I have in mind when I conceive of modulation of automatized predictions, they are insufficient to capture the analytic experience. PLAY as the novel engagement with uncertainty entails a *radical questioning* of the core predictions (S1s) with which one has oriented their lives—a confrontation that entails what Lacan called "subjective destitution" (Verhaeghe, 2019).

Marie Cardinal recounts something her analyst told her when she first consulted him for treatment: "should you agree to come here, it is my duty to warn you...of the risk that psychoanalysis may turn your whole life upside down" (Cardinal, 1975, p. 26). The explosion of flexible engagement with novelty deriving from the particular PLAY I have in mind—perhaps a warped and aberrated PLAY (see Chap. 7)—is not one that seeks to strengthen how well predictions satisfy one's emotional needs. Rather, this particular PLAY, and the PLAYful use of various interventions that aim to *provoke* prediction error, aims at the surplus within the patient's failed predictions.

The modulation of these automatized predictions is no simple matter, because these predictions are the bedrock upon which entire predictive hierarchies have been built for dealing with the contradictions of affective consciousness, the matrices of interpersonal relationships, living in a particular socio-historical context, the personal narrative one has told oneself to constitute an identity, and so on. By uprooting one's most fundamental means of dealing with *jouissance*, the subject in analysis must effectively put into question (and potentially give up) the support of the old predictive network (with its non-declarative and declarative cascades of identity, social position, etc.). Symbolic reduction to S1s and the fundamental fantasy often opposes deeply held beliefs about oneself, one's life, and one's experiences—hence the thrust of techniques like punctuation and scansion *against* the grain of the egoic narrative.

In this sense, the subjective destitution in analysis is a *predictive destitution*. One's core inferences for dealing with the exigencies of life are put into PLAYful abeyance; they are *provoked* and *problematized*. In other words, the PLAY-game of analysis is a deeply difficult game. As Žižek puts it:

> For Lacan, psychoanalysis at its most fundamental is not a theory and technique of treating psychic disturbances, but a theory and practice that confronts individuals with the most radical dimension of human existence. It does not show an individual that way to accommodate him- or herself to the demands of social reality; instead it explains how something like 'reality' constitutes itself in the first place. (Žižek, 2007, p. 3)

To "treat psychic disturbances" could be understood as helping patients better manage and meet their emotional needs. This is one way to read Solms's (2018) conception of analysis. Of course, this *is* an element of analysis, and will perhaps be a greater element of some analyses compared to others.

However, Lacanian psychoanalysis theorizes further (although I suspect that Solms's identification of the premature automatized prediction is already pregnant with this notion). Putting into question one's core, deeply held beliefs (i.e., traversing the fundamental fantasy that has organized *how* one's premature automatized predictions are implemented, such that these automatized predictions might be *modulated* and a different stance toward one's *sinthome* established) "confronts individuals with the most radical dimension of human existence": the fundamental (and contradictory) uncertainty of affective consciousness, one's mode of enjoyment, and the persistent uncertainty throughout one's predictive hierarchies—all so alien, foreign, and disturbing to the self-reflective ego. Beyond adapting to social reality, one goes through this zero-point to question "how something like 'reality' constitutes itself in the first place."

We can be more precise: analysis confronts the subject to question how they constitute their *predictive reality*, the fundamental coordinates (S1s) with which one organizes and infers their world. By PLAYfully modulating predictions, old implementations of (one's mode of) *jouissance* are loosened, made "destitute," and new possibilities emerge—along with the deep recognition that the subject is not all-determined by a predictive network. This PLAY confronts the radical space of structural antagonism and felt uncertainty to open a new stance toward the surplus enjoyment, the *jouissance*, that *is* the extimate, uncanny core of consciousness.

References

Bazan, A., Van de Vijver, G., & Caine, D. (2021). Lacanian neuropsychoanalysis: On the role of language motor dynamics for language processing and for mental constitution. In C. Salas, O. Turnbull, & M. Solms (Eds.), *Clinical studies in Neuropsychoanalysis revisited* (pp. 79–104). Routledge.

Bromberg, P. (1996). Standing in the spaces: The multiplicity of self and the psychoanalytic relationship. *Contemporary Psychoanalysis, 32*(4), 509–535. https://doi.org/10.1080/00107530.1996.10746334

Busiol, D. (Ed.). (2021). *Lacanian psychoanalysis in practice: Insights from fourteen psychoanalysts*. Routledge.

Cardinal, M. (1975/2003). *The words to say it* (P. Goodheart, Trans.). Van-Vactor & Goodheart.

Castonguay, L., & Hill, C. (Eds.). (2012). *Transformation in psychotherapy: Corrective experiences across cognitive behavioral, humanistic, and psychodynamic approaches*. American Psychological Association.

Dall'Aglio, J. (2021). Sex and prediction error, part 3: Provoking prediction error. *Journal of the American Psychoanalytic Association, 69*(4), 743–765. https://doi.org/10.1177/00030651211042059

Dall'Aglio, J. (2023). Extending the theory of premature automatization: The fantasy as an abstract rule in hierarchical cognitive control. *Neuropsychoanalysis, 25*(1), 27–42. https://doi.org/10.1080/15294145.2023.2183888

Dauphin, B. (2023). Precursors of the affective neuroscience project in the writings of Melanie Klein. *Neuropsychoanalysis, 25*(2), 203–216. https://doi.org/10.1080/15294145.2023.2243280

English, C. (2023). *Melanie Klein's narrative of an adult analysis*. Routledge.

Epstein, M. (2019, April). *Buddhism and psychoanalysis: Implications for the psychotherapeutic treatment of trauma*. Paper delivered at the annual Spring Conference for the Rhode Island Association for Psychoanalytic Psychologies in Providence, Rhode Island.

Fink, B. (2011). *Fundamentals of psychoanalytic technique: A Lacanian approach for practitioners*. Norton.

Fonagy, P., Gergely, G., Jurist, E., & Target, M. (2002). *Affect regulation, mentalization, and the development of the self.* Other Press.

Guéguen, P.-G. (2012). Discretion of the analyst in the post-interpretative era. In V. Voruz & B. Wolf (Eds.), *The later Lacan: An introduction* (pp. 10–24). SUNY Press.

Holmes, J., & Nolte, T. (2019). "Surprise" and the Bayesian brain: Implications for psychotherapy theory and practice. *Frontiers in Psychology, 10*, 592. https://doi.org/10.3389/fpsyg.2019.00592

Israely, Y. (2018). *Lacanian treatment: Psychoanalysis for clinicians*. Routledge.

Johnston, A. (2005). *Time driven: Metapsychology and the splitting of the drive*. Northwestern University Press.

Johnston, A. (2019). Lacan's endgame: Philosophy, science, and religion in the final seminars. *Crisis & Critique, 6*(1), 156–187.

Kellman, J. & Radwan, K. (2019, July). *Play celebrates the joy of surprise for evolutionary adaptation*. Paper presented at the twentieth congress of the International Neuropsychoanalysis Association, Brussels, Belgium.

Kellman, J., & Radwan, K. (2022). Towards an expanded neuroscientific understanding of social play. *Neuroscience & Biobehavioral Reviews, 132*, 884–891. https://doi.org/10.1016/j.neubiorev.2021.11.005

Lacan, J. (1956–1957/2021). *The seminar of Jacques Lacan, Book IV: The object relation* (J.-A. Miller, ed., A.R. Price, Trans.). Polity.

Lacan, J. (1964/1978). *The seminar of Jacques Lacan, Book XI: The four fundamental concepts of psychoanalysis* (J.-A. Miller, Ed., A. Sheridan, Trans.). Norton.

Lacan, J. (1975–1976/2018). *The seminar of Jacques Lacan XXIII: The sinthome* (J.-A. Miller, Ed., & A.R. Price, Trans.) Polity.

Lacan, J. (1976a). "Conférences et entretiens dans des universités nord-américaines" [Lectures and interviews at North American universities.]. *Scilicet, 6*(7), 5–63.

Lacan, J. (1976b). Preface to the English-language edition. In J. Lacan *The seminar of Jacques Lacan, Book XI: The four fundamental concepts of psychoanalysis* (J.-A. Miller, Ed., A. Sherian, Trans.) (pp. vii–ix). Norton.

Levenson, H. (1995). *Time-limited dynamic psychotherapy: A guide to clinical practice*. Basic Books.

Milan, B. (2024). *Analyzed by Lacan: A personal account*. (C. E. Landers & C. Vanderwees, Trans.). Bloomsbury.

Miller, J.-A. (2023). *Analysis laid bare*. Libretto Press.

Mobbs, D., Greicius, M., Abdel-Azim, E., Menon, V., & Reiss, A. (2003). Humor modulates the mesolimbic reward centers. *Neuron, 40*, 1041–1048.

Mor-Ofek, H. (2022). Implications of a neuropsychoanalytic formulation in the psychodynamically-oriented psychotherapy of a non-neurological patient. *Neuropsychoanalysis, 24*(2), 159–170. https://doi.org/10.1080/15294145.2022.2127856

Mosri, D. (2023). Report on the 21st congress of the International Neuropsychoanalysis Society, San Juan, Puerto Rico, 2022: "Neuropsychoanalysis: Implications for clinical technique". *Neuropsychoanalysis, 25*(1), 67–86. https://doi.org/10.1080/15294145.2023.2199746

Panksepp, J. (1998). *Affective neuroscience: The foundations of human and animal emotions*. Oxford University Press.

Parr, T., Pazzulo, G., & Friston, K. (2022). *Active inference: The free energy principle in mind, brain, and behavior*. MIT Press.

Rabeyron, T. (2022). Psychoanalytic psychotherapies and the free energy principle. *Frontiers in Human Neuroscience, 16*, 929940. https://doi.org/10.3389/fnhum.2022.929940

Shibata, M., Terasawa, Y., & Umeda, S. (2014). Integration of cognitive and affective networks in humor comprehension. *Neuropsychologia, 65*, 137–145.

Smith, R., & Solms, M. (2018). Examination of the hypothesis that repression is premature automatization: A psychoanalytic case report and discussion. *Neuropsychoanalysis, 20*(1), 47–61. https://doi.org/10.1080/15294145.2018.1473045

Soler, C. (2015). *Lacanian affects: The function of affect in Lacan's work* (B. Fink, Trans.). Routledge.

Solms, M. (2018). The neurobiological underpinnings of psychoanalytic theory and therapy. *Frontiers in Behavioral Neuroscience, 12*, 294. https://doi.org/10.3389/fnbeh.2018.00294

Solms, M. (2021a). A revision of Freud's theory of the biological origin of the Oedipus complex. *Psychoanalytic Quarterly, 90*(4), 555–581. https://doi.org/10.1080/00332828.2021.1984153

Solms, M. (2021b). *The hidden spring: A journey to the source of consciousness*. Profile Books.

Verhaeghe, P. (2019). Lacan's answer to alienation: Separation. *Crisis & Critique, 6*(1), 364–388.

Žižek, S. (2007). *How to read Lacan*. Norton.

Žižek, S., Bonnell, S., Miller, L., & François, M. (2023, July 18). *The game of life* [Video]. IAI Player /video/the-game-of-life

15

Conclusion

Abstract Here I summarize the trajectory of the book. Not only can one bridge Lacanian psychoanalysis with neuropsychoanalysis, affective neuroscience, and computational neuroscience. By putting these concepts into dialogue, new possibilities and ideas emerge in a non-reductive meta-neuropsychology. It is my hope that this book spurs and provokes uncertainty and subsequent theoretical, research, and clinical work.

Keywords Neuroscience • Lacan • Neuropsychoanalysis • Freud • Solms • Sex • Jouissance • Panksepp • Free energy principle • Friston

I hope that, over the course of this book, it is clear how a Lacanian neuropsychoanalysis is not reductive in the theoretical or clinical sense (see Chaps. 2 and 3). The Lacanian registers (real, imaginary, symbolic) and *jouissance* (see Chap. 4) are key concepts for charting a Lacanian take on the brain, computational neuroscience, and Solmsian neuropsychoanalysis. This conceptual bridge is made possible through a *dialogue* between abstract concepts from Lacanian psychoanalysis, Karl Friston's Free

Energy Principle, Jaak Panksepp's affective neuroscience, and Solms's neuropsychoanalysis.

The Free Energy Principle proposes that all self-organizing systems—including brains—aim to minimize uncertainty (see Chap. 5). We do so by forming predictions that minimize prediction error. For Solms, innate emotional systems are special homeostatic hyperprior predictions whose deviations generate affective consciousness as felt uncertainty. The Solmsian child, with limited cognitive capacities, makes the best of a bad job and automatizes predictions (non-declarative action plans) to emotional prediction errors that *do not work* but nevertheless repeat. Hence, patients suffer from affect and seek treatment (Chap. 6).

In a Lacanian neuropsychoanalytic lens (see Chaps. 7, 8 and 9), these concepts take on a different light. Innate affective hyperpriors conflict. Evolutionary inheritance is rift with antagonism. We are conscious *because* of this antagonism—this is the root space of the subject as divided ($). The real of *jouissance* as the surplus prediction error of affective consciousness *emerges* from the contradictions of affective hyperpriors. This overbearing contradiction creates a logical moment of uncertainty of uncertainty, where a decision must made as to which emotional system to prioritize. I distinguish these two moments as J (uncertainty of uncertainty) and J(E) (prioritization of a particular category of uncertainty). With this notation, one is reminded that the prioritized emotional system does not follow a simple homeostatic regime but operates in the space of uncertainty under the repetition-seeking logic of the drive (including the mechanisms of incentive sensitization via dopamine spike-tagging of motoric predictions).

Moreover, the logical moment of J as uncertainty of uncertainty raises the question: why is this particular category of emotion prioritized and not another? The answer to this question—and the landscape of shifting patterns of affective prioritization—is the fundamental fantasy as an abstract rule in hierarchical cognitive control. The fundamental fantasy knots the real of surplus prediction error with the symbolic of non-declarative motoric predictions and the imaginary of declarative predictive implementation and resolution. It creates a *mode of relation* between the subject in the empty space between conflicting hyperpriors ($) and residual prediction error (a) in the field of shared generative models (the Other).

15 Conclusion

Within this mode of relation are *predictions of precision* that govern the selective prioritization of emotional systems [J(E)]. In this way, affects can be ordered in a sequence. J can therefore be informatically displaced, condensed, and substituted through shifting prioritizations of particular categories of uncertainty (Chaps. 10 and 11). Affects thereby operate like signifiers, insofar as they are differentially organized (Chap. 12). More generally, one can designate *levels* of the symbolic separate from language *per se* at different points in the predictive hierarchy from the non-declarative core to the declarative periphery (Chap. 13). Such a perspective opens new sites for clinical intervention that still follow the *logic* of the signifier (see Chap. 14).

Clinically, Lacanian neuropsychoanalysis takes seriously the impossibilities of fully resolving conflicting emotional systems. And our failed solutions to this conflict. We repeat these failures, in and through fantasy, in our dreams, in our slips, in our relationships, and so on. Our failed solutions are our symptom. Our symptom can at once be our cause of suffering but also our most intimate (or extimate) relationship to our fount of consciousness. It is a *mode of enjoyment* (S1-J). It is how we situate ourselves (indefinable at root) to the interpersonal relationships over which we PANIC, the frustrations that bring us RAGE, the novelties we SEEK, the games we PLAY, the experiences we LUST for, the CAREs we share, the threats we FEAR, and all the condensations, displacements, and substitutions we prioritize.

Our symptoms are not solutions to be patched over or replaced with a wiser mode of life. We are only conscious because our homeostases are imperfect. Our failed solutions are to be treasured, to be reinvented, to be vivified anew. In the setting of an analysis—with its encouragement and provocation of surprise, its push to problematize—we can PLAY with this surprise, PLAY with uncertainty, PLAY with our failed solutions. They are repeated, enacted, and rendered precise. They fail again. But they may fail better. And we may PLAY better with them, inventing games out of our uncertainty. And by PLAYing, those symptoms can become less a site of suffering and more a source of creativity. Of novelty. Of surprise. Of metabolizing prediction errors and free energy. And, although we do not know what will follow, that uncertainty becomes precise and precious.

SPRINGER NATURE

GPSR Compliance

The European Union's (EU) General Product Safety Regulation (GPSR) is a set of rules that requires consumer products to be safe and our obligations to ensure this.

If you have any concerns about our products, you can contact us on ProductSafety@springernature.com

In case Publisher is established outside the EU, the EU authorized representative is:

Springer Nature Customer Service Center GmbH
Europaplatz 3
69115 Heidelberg, Germany

The manufacturer's authorised representative in the EU is Springer Nature Customer Service Centre GmbH, Europaplatz 3, 69115 Heidelberg, Germany. If you have any concerns regarding our products, please contact ProductSafety@springernature.com

Printed and bound by CPI Group (UK) Ltd, Croydon, CR0 4YY
10/02/2026
02051061-0002